DISEASES OF FIELD AND HORTICULTURAL CROPS AND THEIR MANAGEMENT

Volume I

(As per ICAR Fifth Dean's Committee Recommendations)

BHUPENDRA SINGH KHARAYAT

Subject Matter Specialist (Plant Protection)
KVK Champawat
G.B. Pant University of Agriculture and Technology
Pantnagar, Uttarakhand (India)

PHI Learning Private Limited

Delhi-110092
2025

*In fond memory of **Shri Asoke K. Ghosh** (October 1942 – February 2024), Founder Chairman and Managing Director of PHI Learning, whose vision endlessly inspires.*

The Legacy Continues....

Published by Pushpita Ghosh, PHI Learning Private Limited, Rimjhim House, 111, Patparganj Industrial Estate, Delhi-110092 and Printed by Syndicate Binders, A-20, Hosiery Complex, Noida, Phase-II Extension, Noida-201305 (N.C.R. Delhi).

₹695.00

DISEASES OF FIELD AND HORTICULTURAL CROPS AND THEIR MANAGEMENT—Volume I
Bhupendra Singh Kharayat

ISBN-978-93-91818-95-1 (Print Book)
ISBN-978-93-91818-96-8 (e-Book)

The export rights of the book are vested solely with the publisher.

For

B.Sc. (Hons) Agriculture Students

Contents

Preface

The book is based on the latest syllabus of ICAR. The Fifth Dean's Committee of ICAR has revised and recommended a new syllabus for B.Sc. (Hons) Agriculture in 2016. The committee has merged the syllabus of disease of field and horticultural crops season-wise, which is a great decision from a practical point of view. *Diseases of Field and Horticultural Crops and their Management*, Volume I, is a core course with 3 (2+1) credit hours and is taught in the fifth semester. There has been a long-felt need for the students and teachers for a lucid, concise, updated and to-the-point book on crop diseases, and their management. The book covers symptoms, etiology, disease cycle and epidemiology, and management of major diseases of the following crops: Field Crops such as Rice, Maize, Sorghum, Bajra, Groundnut, Soybean, Pigeon pea, Finger millet, Black gram and Green gram, Castor, Tobacco and Horticultural Crops such as Guava, Banana, Papaya, Pomegranate, Cruciferous vegetables, Brinjal, Tomato, Okra, Beans, Ginger, Colocasia, Coconut, Tea and Coffee. Some additional diseases are being included such as "Bakanae of Rice", "Papaya Ring Spot" and "Root Knot of Tomato" in this book considering their destructive nature and economic importance, particularly in North India. I hope that this book will be very useful to clear the concept and enhance the understanding of plant disease management of readers as the special emphasis is given in the writing of disease diagnosis, morphology and biology of pathogens, disease cycle, epidemiology, and crop disease management. Additional lists of important crop diseases are also given in the book. The book includes exercises such as MCQs and match the following with answers which are highly useful for ICAR-JRF and other exams such as ICAR-SRF, ICAR-NET, IARI, BHU, AO, AFO, Bank PO, IFS, CSE-UPSC exams, etc. The revised and latest name of the pathogens and recommended practices for crop disease management has been included in the book. Healthy critics and valuable suggestions from students, teachers, and readers will be highly welcomed for the continuous improvement of this book and these will be incorporated in subsequent editions (email: phi@phindia.com or bhupendrakharayat@gmail.com). I hope the book will be highly useful for students, teachers, researchers, plant protection specialists, extension workers, agriculture or horticulture officers, and growers. In the end, I would like to suggest that the students, teachers, and readers must watch the published videos on different crop diseases from my YouTube channel "e-Plant Health" as supporting material for this course or crop disease management.

Bhupendra Singh Kharayat

Acknowledgements

I would like to express my deep gratitude to all those who helped me directly or indirectly during the preparation of this book. First and foremost, I am extremely thankful to all my family members for their constant encouragement and support shown throughout the preparation and publication of these books. I am deeply indebted and highly thankful to all researchers who permitted me to use their work. I am also thankful to my guide Dr. Yogendra Singh (Professor, Plant Pathology, G.B.P.U.A.&T., Pantnagar). I will always be indebted and highly thankful to all my teachers who taught me Plant Pathology at Pantnagar University. I am also grateful to my colleague Dr. Rajani Pant and my mind booster friend Mr. Mahesh Pokharia for their continuous support at odd times. I will always be thankful to all my students who encouraged and inspired me to write this book. My sincere and special thanks to all team members of PHI Learning Private Limited, for extending their support in processing the manuscript and early publication of the book.

Bhupendra Singh Kharayat

1 Rice (*Oryza sativa* L.)

1.1 BLAST

Worldwide, rice blast is one of the most economically devastating crop diseases. Blast disease is one of the major constraints to intensification. Rice blast or rotten neck was first recorded as "rice fever disease" in China by Soong Ying-shin in 1937. In India, the occurrence of this disease was first reported in the Tanjore district of Tamil Nadu. In India, the damage to rice crops due to this disease is estimated as high as 75%. In an outbreak of rice blast disease in Malaysia, yield loss caused by panicle blast was as high as 50%–70%. In the Philippines, rice yield losses range from 50% to 85%. In South and Southeast Asia, 30% losses of the global rice harvest.

Symptoms

The rice blast pathogen fungus produces lesions on all plant parts including leaves, stems, peduncles, panicles, seeds, and even roots. Depending on the site of the symptom rice blast is referred to as leaf blast, collar blast, node blast, and neck blast (Figure 1.1). The colour and shape of lesions vary depending on varietal resistance, environmental conditions, and the age of plants.

Leaf blast: Initial symptoms appear as small bluish flecks about 1–3 mm in diameter. The lesions on the leaf blade are elliptical or spindle-shaped with brown borders and grey centers. Under favourable environmental conditions, lesions enlarge and coalesce eventually killing the leaves.

Collar blast: Appears at the junction of the leaf base and leaf sheath; this can kill the entire leaf.

Figure 1.1 Rice Blast. (A) Leaf blast, (B) Collar blast, (C) Neck blast, (D) Panicle blast.

Node blast: The node of the stem turns blackish and breaks easily; this condition is called node blast. Nodal infection usually appears after heading.

Neck blast or rotten neck or neck rot: Infected neck is girdled by a greyish-brown lesion that makes panicle fall over when infection is severe. Neck becomes shriveled and covered with a grey fluffy mycelium. Neck rot and panicle blast are particularly devastating causing up to 80% yield losses in severe epidemics. The neck rot is the most damaging to the crop and determines the loss.

Panicle blast: Brown lesions appear on the branches of panicles and spikelets. The panicles become white when young neck nodes are invaded. If the infection has occurred much before the grain formation then there will be no grain filling and the panicle will remain erect. The infected pedicels (stalk that bears the flowers) result in the non-production of seeds. This is known as blanking.

Seed lesions: Brown spots or blotches, and occasionally the classic diamond-shaped lesion are often seen on leaves.

Causal Organism and Etiology

Rice blast is caused by *Pyricularia oryzae* (anamorph or asexual stage) or *Magnaporthe oryzae* (teleomorph or sexual stage). It is an ascomycete and classified in the newly constructed family Magnaporthaceae. The fungus produces septate, branched hyaline to slightly coloured mycelia. Conidiophores are produced singly or in groups, are simple, rarely branched, septate, slender, and denticulate, greyish, and show sympodial growth. Conidia are formed singly at the tip of the conidiophores in succession. They are ovate to obclavate, usually 3-celled (2-septate) with a small basal appendage, hyaline to pale olive in colour. Conidia often form appressoria at the tip of the germ tube. Mature conidia are usually three-celled or 2-septate, pyriform (pear-shaped), hyaline or colourless to pale olive, exhibit a basal appendage at the point of attachment to the conidiophore. *Magnaporthe oryzae* is the first fungal pathogen whose genome was sequenced (Dean et al., 2005).

Disease Cycle and Epidemiology

Rice blast is caused by a filamentous ascomycete fungus and is a polycyclic disease spread by asexual spores (conidia) that infect the aboveground tissues of rice plants. The primary sources of inoculum are infected plant residue, collateral hosts, and infected seeds, and in the tropics, the airborne conidia are present throughout the year, enabling stable epidemics to occur year-round. Collateral hosts are sugarcane, *Digitaria marginata*, *Dinebra retroflexa*, *Panicum repens*, *P. proliferum*, *Branchiaria mutica*, *Arundo donax*, *Leersia hexandra*, *Echinochloa crus-gali*, *Digitaria sanguinalis*, *Stenotaphrum secondratum*, and *Eremochloa ophiuroides*. The infected seeds can produce the conidia for several weeks, i.e. enough time for the infection of the seedlings. It can also sporulate on seed lesions. The disease cycle begins when airborne conidia land on a rice plant and adhere to the host surface through the production of spore tip mucilage. This is followed by a series of developmental steps: *germination*, *germ tube growth*, *formation of an appressorium*, *emergence of a penetration peg*, and *subsequent invasive growth in the host*. After landing on rice leaves, conidia germinate in presence of a thin water film (rain, dew, guttation, drops, etc.) by producing a germ tube which later forms appressorium. The appressorium is a melanized structure, and from it, an infection or penetration peg develops (at the base of appressorium) which penetrates the host tissue. The melanized appressorium generates hydrostatic turgor pressure of up to 8 MPa, which is a sufficient mechanical force to penetrate directly through the leaf cuticle and enter the plant epidermal cells with the help of a penetration peg. In susceptible hosts, the fungus ramifies between and within cells, probably spreading from one plant cell to another through clusters of plasmodesmata, called pit fields. Within a few days of infections, lesions appear. Conidia produced on the lesions cause secondary infections which repeat several times, this process is known as the polycyclic disease cycle. Under favourable conditions, one cycle (conidia germination to conidia production) per week can occur. Melanised macroconidia are then produced on conidiophores that protrude from lesions, with up to about 20,000 spores produced from one lesion on a leaf, and 60,000 on a single spikelet in one night. The fungus is hemibiotrophic, initially feeding on living cells (biotrophic stage) but subsequently causing cell death (necrotrophic stage). Spore release is triggered by a 1 to 2 hour period of darkness. The fungus can sporulate repeatedly for around 20 days. The sporulation, infection, disease spread, and development are favoured by moist conditions, long periods of leaf wetness (for 6–8 hours), relative humidity ranging from 80%–100%, and temperatures ranging from 25°C–30°C. Germination of conidia occurs between 10°C–33°C, the optimum being 25°C–28°C. Drought stress and the application of excess nitrogen increase the susceptibility of rice to the blast pathogen. A forecasting model called *Epi-Bla* has been developed in India.

Management

Rice blast has become more difficult to manage because of the pathogen's ability to survive and multiply in harsh environmental conditions and easily spread to new fields. The economic threshold level (ETL) for the blast is 3–5 lesions per leaf early to late tillering or 2–5 neck 2 infected plants per meter at panicle initiation to booting.

Cultural Methods

- **Seeds or planting material:** Use disease-free seeds. Using high-quality and disease-free seeds is always highly recommended because infested seeds left on the soil surface provide inoculum from which epidemics develop.
- **Crop rotation:** If infected crop residue is left in the field then crop rotation can be the simple and effective technique that is highly recommended. Otherwise, it is an airborne disease.
- **Spacing:** Avoid close spacing of seedlings in the main field.
- **Field sanitation:** Destroy stubbles, weeds, and collateral hosts in and around rice fields and bunds.
- **Planting time:** Adopting early sowing of *Kharif* rice in blast endemic areas.
- **Optimum use of fertilizers:**
  - Overuse of nitrogen fertilizers increases the severity of rice blast in the fields.
  - Split and judicious application of nitrogenous fertilizers.
  - Neither phosphorus nor potassium has any effect on blast incidence in susceptible varieties.
  - Application of silicon fertilizer to rice improves fitness in nature, increase productivity and resistance to blast, brown spot, and sheath through mechanical barriers, accumulation of phenolics and phytoalexins, and activation of pathogenesis-related (PR) genes.
  - Apply a sufficient amount of farmyard manure and green manure during land preparation.
- **Water level:** Low water level increases the rice blast disease.
- **Weeding:** Remove or destroy the weed host or collateral host of blast pathogen from the field.

Host Resistance

- Use resistant or tolerant rice varieties for cultivation such as Rashi, Vikas, Krishna, Hamsa, Tulasi, IR64, Aditya, Swaranadhan, Himalaya 1, Himalaya 2, Himalaya 2216, Pant Dhan 4, Pant Dhan 10, HKR 228, PNR 519, Tikkana, Vajram, Vasistha, Vikramarya, Vandana, Hemavathi, Madhuri, Kanchana, Swetha, Karjat 2, Ratnagiri-1,

Sugandha, Indira, Indravati, Moti, Pratap, Rudra, ADT 36, ADT 39, CO 45, TKM 10, IR 24, Ratna, Suraksha, Apurva, Chaitanya, Gautami, MTU 9993, Phalguna, Penna, Pinakini, Raja Vadlu, Sagar Samba, Simhapuri, Sriranga, Swarnamukhi, Ajaya, IR 36, IR 64, Jagannath, Jaya, Vijetha, Cottondora Sannalu, Prabhat, Nellore Mahsuri, Krishna Hamsa, Pusa 205, and Pusa Basmati 1.

Biological Control

- Use antagonistic bioagents like *Pseudomonas fluorescens* for seed treatment (@ 10 g/kg of seed) and foliar sprays.
- Seedling root dipping in *Pseudomonas fluorescens* (liquid formulation 2.0 × 10^8 CFU, 10 ml/litre of water; powder formulation 10 g/L of water) for 30 minutes.

Chemical Control

- Treat the seeds with Carbendazim @ 1.0 gram (g) or a mixture of Carbendazim 25% + Mancozeb 50%–75% WP @ 2.0 gram (g)/kilogram (kg) seed per litre (L) of water as a wet seed treatment.
- Spray Carbendazim 50% WP @ 250–500 gram (g)/hectare (ha) or Isoprothiolan 40% EC @ 750 ml/ha or Tricyclozole 75% WP @ 300–400 g/ha or Tricylazole 70% WG @ 300 g/ha in the nursery and vegetative stage, panicle initiation to booting stage.
- Spray Ediphenphos 50% EC@500–600 ml/ha or Isoprothiolan 40% EC@750 ml/ha or Tricyclazole 75% WP @ 300–400 g/ha or Kasugamycin @ 2.5 ml/L, during the flowering stage.
- Spray Indofil Z-78, 75 WP (Zineb) @ 500 g per acre in 200 litres of water, at the maximum tillering and ear-emergence stages. The first spray should be given at the boot stage and the second after 15 days.

1.2 BROWN SPOT

It is also called *nai-yake*, i.e. seedling blight, sesame leaf spot and Helminthosporiosis. In India, the brown spot of rice was first time reported by Sundararaman from Madras in 1919. In India, the disease is more prevalent in all rice-growing areas, especially in heavy monsoon areas in West Bengal, Assam, Kerala, Tamil Nadu, Uttarakhand and Eastern parts of U.P. Brown spot is a fungal disease that can infect both seedlings and mature plants. The disease causes blight on seedlings, which are grown from heavily infected seeds, and can cause 10%–58% seedling mortality. The disease is of historic significance and considered to be the major factor contributing to the *Great Bengal Famine* in resulting yield losses of 50%–90% and causing the death of 2 million people. Brown spot is still a devastating disease to rice.

Symptoms

Small brown lesions, ellipsoidal or oval, most conspicuous on the aerial parts such as coleoptiles, leaf blades, leaf sheaths, panicle branches, glumes and spikelets. Initially, the spots are surrounded by a golden-yellow halo. Necrotic lesions with a light-brown to greyish center surrounded by a dark-reddish brown margin appear on foliage at maturity. In the advance stage of severe infection, the lesions coalesce often leading to the field appearance as burnt and scorched and to the development of "pecky rice", i.e. grains spotting, shriveling, and discolouration of the seeds. The fungus produces a toxin, ophiobolin that causes necrotic symptoms in plants. The spots of brown spot disease can be confused with those of blast but are easily distinguished. The brown spot lesions (Figure 1.2) are oval to circular, while blast lesions are spindle-shaped.

Figure 1.2 Brown Spot of Rice. (A) Rice plants infected with brown spot disease, (B) Close up of brown spot symptoms.

Causal Organism and Etiology

Brown spot of rice caused by *Bipolaris oryzae* (=*Helminthosporium oryzae*) or (telemorph=*Cochliobolus miyabeanus*). The fungus produces inter and intra-cellular mycelium, which develops as greyish-brown to dark-brown mat on the infected tissues. The conidiophores are thick, erect, geniculate, branched or unbranched, dark olivaceous at the base, and lighter towards tip, and emerge through stomata in clusters of 3–5. The conidia are hyaline, cylindrical, or slightly curved and wider in the middle, 3–10 septate and bear a scar, indicating the point of attachment on the conidiophore. In the perfect stage, *Cochliobolus miyabeanus* produces perithecia in culture appears in clusters and is globose, pseudoparenchymatous, black, and with an ostilar beak. Asci are cylindrical to long, fusiform, slightly curved, and contain mostly 4–6 filiform ascospores.

Disease Cycle and Epidemiology

The major sources of inoculum for brown spot disease in the field are infected seeds, volunteer rice, infected rice debris, several weed hosts, and soil. The fungus can survive in the seed for more than four years and give rise to infected seedlings after germination. However, the pathogen does not survive for long in infected plant debris and natural soil. The fungus naturally causes infection in as many as 20 different wild species of *Oryzae*. A few collateral hosts like *Digitaria sanguinalis, Leersia hexandra, Panicum maximum* (guinea grass), *P. repens* (torpedo grass), *Echinochloa colona, Pennisetum typhoides, Pennisetum purpureum* (elephant grass), *Setaria italica* and *Cynodon dactylon*, are another source of primary inoculum. The conidia carried through infected rice seeds germinate on the emerging seedlings and produce several generations of asexual spores (conidia), as the plants grow. The conidia are dispersed by wind and cause infection to the new crop and thus serve as the source of secondary infection. The airborne conidia germinate on the leaf surface. The optimum temperature for conidia germination is 25°C to 30°C and humidity is 90% or above and leaf wetness is required for 8–24 hours. Within a few hours, the tip of the germ tube swells to form a lobed or branched appressorium from which penetration peg develops and penetrates through the epidermis or stomata. Infection through stomata without the formation of appressoria may also occur. The fungus grows beneath the stomata and remains confined to the parenchyma. The lesions first appear about 21 days after infection. The fungus sporulates and forms conidiophores that emerge out in groups through the stomata. The pathogen can spread fast and may cause devastating epidemics. The factors favouring the rapid growth of fungus and brown spot infection are leaf wetness for 8–24 hours, cloudy weather, warm days with temperature ranging from 25°C–30°C and high humidity (86% to 100%), water stress, low sunshine hours, susceptibility of the host, which is maximum during maturity period. Other factors favouring the disease development are rice suffering from N, K, or P deficiency, the presence of infected seeds, volunteer rice, rice debris, and several weeds, poorly drained or abnormal soils.

Management

Cultural Methods

- Use of seeds from a disease-free crop as the fungus can survive on seeds for up to 4 years
- Improve soil fertility especially silica-based fertilizer.
- Remove infected rice straw and weed hosts such as *Echinochloa* spp., *Leersia hexandra.*
- Apply well-balanced nutrients to the soil.
- Avoid water stress conditions.
- Clean all rice debris and weeds from fields that are the sources of inoculum.
- Remove alternate and collateral hosts.

Physical Control

- As the fungus is seed transmitted, so treat seeds with hot water (53°C–54°C) for 10–12 minutes before planting, to control primary infection at the seedling stage. To increase the effectiveness of treatment, pre-soak seeds in cold water for eight hours.

Host Resistance

- Grow resistant varieties like Bala, BAM 10, IR-20, Jaya, Ratna, Tellahamsa, ADT 44, PY 4, CORH 1, CO 44, Cauvery, Bhavani, TPS 4, Dhanu, and Kakatiya.

Biological Control

- Treat the seed treatment with *Pseudomonas fluorescens* @ 10 g/kg of seed followed by seedling dip @ 2.5 kg/ha dissolved in 100 litres and dipping for 30 minutes.

Chemical Control

- Seed soak or seed treatment with Captan or Thiram or Chitosan or Carbendazim or Mancozeb @ 2 g/kg of seed to reduce seedling infection.
- Seed treatment with Tricyclazole followed by spraying of Mancozeb + Tricyclazole at tillering and late booting stages gave good control of the disease.
- Give 3–4 sprays of fungicides like Dithane Z-78 or Dithane M-45 @ 2.5 g/L water at 10 days intervals just before the appearance of initial symptoms of the disease.
- Give two sprays of Tebuconazole 25 EC @ 200 ml or Tilt (Propiconazole) 25 EC @ 200 ml or Indofil Z-78 (Zineb) @ 500 g in 200 L of water/acre. The first spray should be given at the boot stage and the second after 15 days.
- Spray Mancozeb (2.0 g/L) or Edifenphos (1 ml/L) 2 to 3 times at 10–15 day intervals. Spray preferably during the early hours or afternoon at flowering and post-flowering stages.
- Griseofulvin, Nystatin, Aureofungin, and similar antibiotics have been found effective in preventing primary seedling infection.

1.3 SHEATH BLIGHT

Rice sheath blight is one of the most economically significant rice diseases worldwide. The disease is also called *snake skin disease, mosaic foot stalk, oriental sheath and leaf blight* and *rotten foot stalk* because of its special disease symptoms. Sheath blight was first reported from Japan in 1910 by Miyake, but was first noted by Yano in 1901. But in India, Paracer and Chahal reported this disease from Gurdaspur (Punjab) in 1963. Yield losses of up to 50% have been reported under the most conducive environments. This is also a very devastating disease in India. It causes severe loss approximately 40%–45% under favourable environmental conditions. Sheath blight is one of the highly destructive diseases of rice, considered a globally significant one, and second-most prevalent to the blast disease.

Symptoms

Leaf sheath: Early symptoms develop as circular, oval, or ellipsoid water-soaked greenish-grey lesions, usually 1–3 cm long, on the leaf sheaths at or just above the water line in lowland fields (Figure 1.3A). As the disease progress, they enlarge and may coalesce forming larger lesions with irregular elongated outline and greyish-white center with dark purplish or dark brown borders. Eventually, individual lesions coalesce, even covering the entire sheath and consequently, the sheath turn yellow, wilt, and even rot and dies.

Leaf and leaf blades: Usually irregular lesions, often with grey-white centers and dark green, brown, or yellow-orange margins, as they grow older, appear on leaves. The lesions can develop extensively and coalesce on partial or whole leaf blades, which may produce *a rattlesnake skin* *pattern* (Figure 1.3C). Under wet and humid conditions, microscopic runner hyphae (white, web-like hyphae) of the fungus grow on infected leaves and sheaths which spread the disease rapidly, from leaf to leaf, and plant to plant. Microscopic runner hyphae and sclerotia are two major signs of fungal infection. White (young) to brown (older) sclerotia of the fungus appears on the infected sheaths, leaves, and leaf blades.

Panicle: Under favourable conditions, the fungal infection can extends rapidly up to the panicle. The panicle infection results in the production of sterile or partially filled, discoloured seeds with brownish-black spots or black to ashy grey patches (Figure 1.3B). In case of severe infection, white mycelial growth and sclerotia can cover the infected ears.

Figure 1.3 Sheath Blight of Rice. Symptoms on sheath (A), ear (B), and leaves (C).

Casual Organism and Etiology

Rice sheath blight is caused by *Rhizoctonia solani* (teleomorph: *Thanatephorus cucumeris*). *Thanatephorus cucumeris*, the sexual stage of the causal agent of RSB, belongs to the Corticiaceae family in the Homenomycetales order of the Basidiomycetes class. *R. solani* has been assigned into 14 anastomosis groups (AG1 to AG13, AGB1) with high genetic diversity based on their compatibility for hyphal fusion with known tester isolates. In India, the sheath blight of rice is caused by anastomosis group 1 (AG-1 IA) of fungus having 3–16 nuclei per cell.

Disease Cycle and Epidemiology

Rhizoctonia solani survives through sclerotia and mycelia in infected plant debris/straw, stubbles or ratoons, and seed and also through weed hosts in tropical environments while in temperate regions, the primary source of inoculum is sclerotia. The pathogen is generally soil and waterborne, infectious to a broad of range over 250 plant species belonging to about 32 taxonomic families. A typical inoculum of the disease is either sclerotia or runner hyphae from the infected plants. During the cropping season or at harvest, sclerotia fall on the ground and serve as survival structures from one cropping season to other. These sclerotia may remain dormant for over the years in the soil and stubble and can re-infect healthy rice plants in the subsequent crop season. Basidiospores of the fungus rarely serve as inoculum. *R. solani* is a monocyclic fungus as it does not produce any kind of asexual spores. The life cycle of *R. solani* begins when sclerotia or a piece of infected plant debris floating on the water surface come in contact with newly transplanted seedlings, germinates by producing mycelia and causes the infection. Mycelia or runner hyphae further grow and produce infection cushions (aggregates of complex hyphae) or lobate appressoria or both on the plant surface. Infection peg develops from appressoria or infection cushions which penetrate through the cuticle or stomata or wounds. The mycelia then move upward along the sheaths and leaves of rice plants, ultimately resulting in damage to the sheaths, leaves, and grains. The fungus spreads in the field by growing its runner hyphae from tiller to tiller, from leaf to leaf, and from plant to plant, resulting in a circular pattern of damage. The life cycle ends with the formation of overwintering structures, sclerotia, on the sheaths, leaves, seeds, and in soils. Sclerotia are loosely attached to infected plant parts, and may easily dislodge at maturity and can float due their buoyancy over long distances in the field with irrigation water and aid in the spreading of the disease. Seed infection and transmission of the pathogen from seed to seedlings in the form of brownish-black to blackish lesions on coleoptile, radical, leaf, and sheath. Sheath blight also infects weed hosts and causes similar symptoms. In addition, the fungus also infects other crops including corn, sorghum, soybeans, and sugarcane, increasing the inoculum in the soil. Factors that favour the infection spreads most quickly and development of sheath blight include high temperature (28°C–32°C), relative humidity of crop canopy (85%–100%), high levels of nitrogen fertilizer, use of highly susceptible semi-dwarf rice varieties, short rotations with non-host crops, overuse of nitrogen fertilizer, and dense plant stands that create favourable microenvironments in the canopy for the pathogen.

Management

Cultural Methods

- Use healthy seeds of rice.
- Avoid excessive use of fertilizers.
- Avoid the flow of irrigation water from infected fields to healthy fields.
- Deep ploughing in summer and burning the stubble.
- Follow crop rotation regularly.
- Uproot or remove all weeds present on bunds from or nearby the rice field.

- Remove the weeds from bunds on regular basis.
- Drain off standing water from the infected field as the pathogen can spread with water current from one field to other.
- Transplant rice far from bunds about 50–60 cm so that the weeds on bunds cannot touch the rice plants.
- Drain rice fields relatively early in the cropping season to reduce sheath blight epidemics.
- Avoid close planting. Improve canopy architecture by reducing the seeding rate or providing wider plant spacing.
- Apply organic amendments.
- Apply FYM 12.5 t/ha or green manure 6.25 t/ha to promote antagonistic microflora.

Biological Control

- Treat the rice seed with bioagents like *Trichoderma harzianum* or *Pseudomonas fluorescens* @ 10 g/kg of seed, before sowing in the nursery.
- Seed treatment with *Pseudomonas fluorescens* @ 10 g/kg of seed followed by seedling dip @ 2.5 kg/ha dissolved in 100 litres and dipping for 30 minutes.
- Soil application of *P. fluorescens* @ 2.5 kg/ha after 30 days of transplanting (the product should be mixed with 50 kg of FYM or sand and then apply).
- Foliar spray of *Pseudomonas fluorescens* @ 0.2% commencing from 45 days after transplanting at 10 days intervals for 3 times depending upon the intensity of disease.
- Spray *Trichoderma harzianum* + *Pseudomonas fluorescens* @ 10 g/L of water.
- Soil application of *Pseudomonas fluorescens* (2.5 kg/ha with 100 kg of FYM at field preparation).
- Foliar application of *Pseudomonas fluorescens* @ 2.5 kg/ha with 500 L of water.

Host Resistance

- Grow resistant varieties like PR 108, Bhudeb Dinesh, Jogan, Mandira, Nalini, Neeraj, and Sabita.
- Grow short duration varieties such as IET-1410, Ratna, and PC-19.

Chemical Control

- Treat seed with Carbendazim @ 2 g/kg of seed.
- Spray Validamycin 3% L @ 2000 ml/ha or Hexaconazole 5% EC @1000 ml/ha or Propiconazole 25% EC @ 750 ml/ha or Propiconazole 10.7% + Tricyclazole 34.2% SE @ 500 ml/ha at vegetative growth, panicle initiation to booting and flowering stage.
- Spray Hexaconazole 5 EC @ 2.0 ml or Validamycin 3 L @ 2.0 ml or Propiconazole 25 EC @ 1.0 ml or Trifloxystrobin 25% + Tebuconazole 50% WG @ 0.4 g or Hexaconazole 75 WG @ 0.13 g/L or Azoxystrobin 25 SC @ 1.0 ml/L of water, in heavily infested soils.

1.4 FALSE SMUT

False smut disease, also known as *orange or green smut* or *haldi rog* or *Laxmi rog*. It had been recognised as a symbol of a bumper harvest and in some parts of Southern India it is popularly known as *Laxmi* (Goddess of wealth and prosperity) disease. It was first reported from the Tirunelveli district of Tamil Nadu (India) in 1878. It is not a true smut, but is classified in a different division of the Kingdom Fungi, the Ascomycota, while true smuts belong to Basidiomycota. Because of the sporadic nature of the disease, emphasis was not given in the past for managing this disease in India. It has been reported that in India, the percentage of false smut–infected tillers ranged from 5%–85%, and the disease caused about 0.2%–49% yield loss depending on rice varieties and disease intensity. Besides the direct economic loss to the rice-growing farmers, the pathogen also produces poisonous mycotoxin known as *ustiloxin* in false smut balls. Ustiloxins are cyclic peptides, which are toxic to plants, humans and animals and interfere with microtubule function, thereby inhibiting mitosis. The disease not only causes a significant yield loss but also reduces grain quality by generating a variety of mycotoxins, such as ustiloxins and ustilaginoidins, which are toxic to both humans and animals.

Symptoms

The typical symptom of rice false smut disease is the formation of a white fungal mass protruding from the inner space of a spikelet at the early infection stage, followed by the replacement of rice grains with false smut balls, which are at first yellowish orange to green, and finally turn a greenish-black (Figure 1.4). Immature yellow spores are covered by a thin cream-coloured membrane which later ruptures at the time of maturity of the spores. The outer layer of the mature smut ball consists of numerous chlamydospores and is often covered by hard, black horseshoe-shaped sclerotia. In most cases, not all spikelets of a panicle are affected, but spikelets neighboring smut balls are often unfilled. Spores contaminate adjacent grain. False smut ball is the only visible symptom of this disease identified so far.

Figure 1.4 False Smut of Rice.

Causal Organism and Etiology

False smut of rice is caused by *Ustilaginoidea virens* (teleomorph: *Villosiclava virens*). It is a nonobligate biotrophic, ascomycete fungal pathogen. The pathogen belongs to the kingdom: Fungi, phylum: Ascomycota, class: Ascomycetes, subclass: Sordariomycetes, order: Hypocreales, family: Clavicipitaceae, genus: *Villosiclava* and species: *virens* and its anamorphic stage are *Ustilaginoidea virens.* The conidia are round to elliptical. Upon maturation or under unfavourable conditions, conidia may develop into rounded chlamydospores with prominent irregularly curved spines on the surface. Sclerotia are dark brown to black horseshoe-shaped and irregular oblong or flat, ranging from 2–20 mm which after germination produce tiny toadstool like fruiting bodies ascostromata containing perithecia that produces ascospores. Sclerotia survive on or in the soil for up to 12 months.

Disease Cycle and Epidemiology

False smut is a soil, seed, and airborne fungal disease. Besides infecting of rice, *U. virens* also infect several common weeds rice fields such as *Digitaria marginata, Panicum trypheron, Imperata cylindrical,* and *Echinochloa crus-galli.* These weed species may also the support epiphytic growth of *U. virens* and play certain roles in the disease cycle. In temperate regions, the fungus survives the winter through sclerotia as well as through chlamydospores. The life cycle of *U. virens* involves both sexual and asexual stages. Both sexual and asexual spores can infect the rice spikelet and convert a rice grain into a smut ball. In the sexual cycle, sclerotia, which are induced by low temperatures, form on the surfaces of false smut balls. Under favourable condition such as optimum moisture, humidity, light, and temperature, the sclerotia germinate and produce special yellow fruiting bodies, i.e. stromata full of perithecia at the tip. The perithecia contain a large number of asci embracing eight immature ascospores. In normal conditions, a single mature sclerotium can produce millions of ascospores. The ascospores produce secondary conidia to contribute to primary infections of rice. During the asexual cycle, thick-walled chlamydospores form on the surfaces of the false smut balls which may contaminate rice seeds and/or spread by air currents and rain splash in paddy fields, where they survive during the winter. Under high humidity and watery conditions, some chlamydospores may germinate to form hyphae to generate a large number of secondary conidia, which may infect rice flowers at the late booting stage. In general, *U. virens* hyphae spread over the outer surface of the spikelets and extend into the inner space of the spikelets through the gap between the lemma and palea to infect the stamen filaments, and possibly the stigma or lodicules, without haustorium or appressorium. Chlamydospores are important in secondary infection which is a major part of the disease cycle. Therefore, both sclerotia and chlamydospores have the potential to be primary inoculum in the field. After *U. virens* infection at the late booting stage, rice spikelets are usually transformed into false smut balls covered with powdery, dark-green chlamydospores after heading. The disease occurs in areas with high relative humidity (more than 90%) and high temperature (25°C–35°C), during flowering. Rain, high humidity, and soils with high nitrogen content also favour disease development and wind can spread the fungal spores from plant to plant.

Management

Cultural Methods

- Use high-quality disease free certified seeds. Seeds contaminated with fungal spores and sclerotia may serve as inoculum for further spread and development.
- Follow crop rotation with non-host crops.
- Remove alternative hosts, including grassy weeds, especially common barnyard grass (*Echinochloa crus-galli*) and Jungle rice (*Echinochloa colona*).
- Keep the field clean, remove infected seeds, panicles, plant debris after harvest, and remove the alternate host.
- Early planted crop has less smut ball than the late planted crop.
- Deep plough to at least 6 inches.
- Practice sun drying the field during the summer.
- Regular monitoring of disease incidence is very essential to take necessary proactive measures.
- Balanced use of fertilizers and micro-nutrients as per local recommendations.
- Reduce humidity levels through alternate wetting and drying of fields rather than permanently flooding the fields.
- At the time of harvesting diseased plants should be removed and destroyed, so that sclerotia do not fall in the field. This will reduce the primary inoculum for the next season.

Physical Control

- Hot water treatment of seeds at 52°C for 10 minutes.

Biological Control

- Treat the seeds with *Pseudomonas fluorescens* 10 g/L of water for 30 minutes or *Trichoderma viride* @ 5–10 g/kg of seeds protects from soilborne and seedborne diseases.
- Dip the seedlings with the solution of *P. fluorescens* @ 5 g/L of water for 20 minutes before transplanting.
- Spray *P. fluorescens* @ 5 g/L of water at an interval of 15–20 days after transplanting.
- Apply Neem cake @ 150 kg/ha.

Chemical Control

- Spray only once a fungicide such as Carbendazim or Hexaconazol @ 0.1%, 5–7 days before flowering stage.
- Spray Copper oxychloride @ 0.25% at booting stage followed by a second spray of Propiconazole @ 0.1% at 10 days after the first spray.

1.5 BAKANAE OR FOOLISH SEEDLING

Bakanae is one of the oldest known diseases of rice in Asia. Bakanae disease of rice was known to be present from 1828 in Japan, but it was first described by Shotaro Hori (a Japanese researcher) in 1898. In India, the disease was first reported by Thomas in 1931 from Andhra Pradesh. As every plant biology student learns, gibberellin (GA) was first isolated in Japan in the 1930s by Japanese farmers from a fungus that causes serious disease in rice, called *foolish seedling disease*. The purified active compound was named for the fungus, an ascomycete called *Gibberella fujikuroi* (sexual stage). The name bakanae (Japanese word) means *bad* or *thin noodle seedling*, *stupid rice crop* or *foolish seedlings* because of specific elongation symptoms. In different countries, the disease is known by various other names also such as in French Equatorial as *foot rot* and *bakanae*, in China as *white stalk* and *Fusariosis*, in British Guiana as *palay lalake* (man rice), in Japan as *otoke nae* or *bakanae-byo* (male seedling, elongation disease) and in India as *foolish plant* or *foot rot* or *jhanda rog*. Bakanae is emerging as one of the most serious diseases of rice in India. In various rice growing countries, significant yield losses caused by the disease can range from 50% to more than 70%. Low plant survival and high spikelet sterility may account for yield losses of up to 3.0%–95% in India, 40% in Nepal, 50% in Japan, 75% in Iran, 28.8% in Korea, and 6.7%–58.0% in Pakistan.

Symptoms

The typical symptoms of bakanae disease include seedling blight, root rot, crown rot, stunting, hypertrophy effect or excessive elongation (the most classical symptom), foot rot, fewer tillers with pale green flag leaves, seedlings rot, grain sterility, grain discolouration and poor seed quality under different climatic conditions around the world (Figure 1.5). The disease occurs in the nursery and main fields. Infected seedlings have lesions on the roots and can die before transplanting or immediately after. Infected seed results in poor seedling emergence and gaps in the rows. Generally infected plants die. Surviving seedlings are slightly chlorotic, thin, excessively elongated with fewer tillers, and may wilt and die later on. Surviving seedlings that reach maturity are taller due to abnormal elongation of internodes and bear only empty panicles (glumes) and white ears. Pink-white fungal growth (mycelium and spores) develops at the soil level on infected plants that move up the stems. The nodes on infected plants may be pink to purple beneath the leaf sheaths. Mycelial masses with sporulation are present inside the internodes. Adventitious roots develop on the first two or three nodes (lower nodes) above the ground. Root rot results from soilborne infections. The rapid and excessive elongation of infected plants is attributed to gibberellins (plant growth regulator) and stunting to fusaric acid (toxin) produced by the pathogen.

Causal Organism and Etiology

Bakanae disease of rice is caused by one or more seed-borne *Fusarium* species, mainly *F. fujikuroi* (sexual stage: *Gibberella fujikuroi*). To date, four *Fusarium* species are associated with bakanae disease of rice, including *Fusarium fujikuroi*, *Fusarium andiyazi*, *Fusarium verticillioides* and *Fusarium proliferatum* in the *Gibberella fujikuroi* species complex. *F. fujikuroi* is the most virulent and widespread species among these species complex and produce gibberellin which increases disease severity. *F. fujikuroi* is hemibiotroph, which causes initial infection on a living host as a biotroph and later switches to a necrotrophic life style (kill the host cells). The pathogen produces sexual (ascospores) and asexual spores (macro-conidia and micro-conidia). Ascospores are piston shaped, cylindrical, and flattened. Sporodochia are normally absent but, if present they are pale orange and contain macroconidia. Microconidia are hyaline, and oval or club-shaped with flattened bases with 0 to 1 septa (mostly single-celled) while macroconidia are slightly sickle-shaped, narrow at both ends, and 2–5 celled. The pathogen survives by producing sclerotia (dark blue and spherical). False heads and chains of short to medium length are produced from polyphialides, which may proliferate and often form monophialides, whereas chlamydospores are absent. *F. fujikuroi* is also known to produce fusaric acid, beauvericin, gibberellic acid, and other secondary metabolites. A few strain of *F. fujikuroi* also known to produce fumonisins B1, B2, and B3 instead of gibberellic acid.

Figure 1.5 Bakanae or Foolish Seedling of Rice. (A) Foot rot symptoms, (B) White-pink fungal growth on stem, (C) Excessive elongation of rice plants compare to healthy, (D) White ear formation (empty panicles).

Host Range

- **Primary hosts:** Rice, maize, barley, sorghum, sugarcane, wheat, pine, rye, and asparagus.
- **Alternate hosts:** Tomato, cowpea, banana, subabool, proso millet, early water grass, and barnyard grass.

Disease Cycle and Epidemiology

Bakanae is a monocyclic, seed, and soilborne disease. Seedborne inoculum is a more significant source and provides initial foci for primary infection. The pathogen dispersed predominantly with infected seeds and contaminated seeds. Long-distance spread is on and in infected seed. The pathogen survives as spores on the seed coat and as thick-walled hyphae or macroconidia in infected crop residue in the soil. The fungus remains viable in seed for about 1–2 years, and about 1 year in the soil. Ascospores and conidia adhering to the seed germinate and infect seedlings through roots and crowns. Infection may also take place through spores and mycelium that are left in the water used for soaking seeds to stimulate germination before sowing. Infected seed results in elongated plants (bakanae) while soilborne infection leads to root rot. From the roots, fungus grows systemically within the plant and sporulates. The conidia are disseminated by wind and water causing new (secondary) infections in the rice field. The sporulation of fungi coincides with the flowering and harvesting stage of the crop resulting in seed infections and contaminations. Contamination of panicles is caused by airborne conidia (asexual spores) and ascospores (sexual spores) produced on leaf sheaths, stems, and diseased or dead stubbles. *F. fujikuroi* infects rice flowers and transmits the next plant generation via seeds. Bakanae infection is favoured by dryland cultivation and high temperatures (above 30°C) and relative humidity (above 90%). The optimum temperature for pathogen infection is 27°C–30°C and for disease development 35°C which is also highly favourable for seedling growth. Lower temperatures and very dry soil favour root rot.

Management

Cultural Methods

- Use of clean (non-infected) rice seeds is important to prevent bakanae disease as it is a seedborne disease.
- Use salt water to separate light weight, infected seeds during soaking.
- Destroy the crop residue.
- Late planting of rice seedlings minimised the incidence of bakanae disease.
- Follow crop rotation.

Physical Methods

- Thermal seed treatment (hot water immersion) is the most common and wide method to limit bakanae disease occurrence. Immersion of infected seeds in hot water at 58°C or 60°C for 10–20 minutes has shown a similar disinfection efficacy to the conventional chemical treatment.
- Sterilization of infected seeds with heated acidic electrolyzed water at 50°C for about 10 to 20 minutes can minimise the disease.

Host Resistance

- Scented rice genotypes such as C4-64 (green base), Karjat × 13-21 are resistant and nonscented rice genotype such as IR 58109-109-1-1-3, BR 1257-31-1-1, BR 1067-84-1-3-2-1, and BR 4363-8-11-4-9 are highly resistant to bakanae disease. Other genotypes such as Chandana, C101A51, Athad apunnu, IR 58025B, PAU 201, Panchami, Varun Dhan, and Pusa 1342 as highly resistant, whereas Himju, BPT 5204, Suphala, and Peeli Badam have been found as resistant to bakanae disease.

Biological Control

- Dip roots of rice seedlings in spore suspension (10 g/L) of *Talaromyces flavus* or *Trichoderma harzianum.*

Chemical Control

- Treat the seed with 0.3% Sodium hypochlorite (NaOCl) solution for 24 hours.
- Treat the seed with a fungicide such as Benomyl (at the rate of 1%–2% of seed weight) for dry seed coating.
- Soak the seed in 0.1%–0.2% solution of fungicides such as Benomyl, Triflumizole, Propiconazole, and Prochloraz for one hour or in 0.05% solution of a combination of Thiram + Benomyl or Carbendazim for 5 hours.

1.6 BACTERIAL LEAF BLIGHT

Bacterial leaf blight (BLB) of rice is one of the most devastating diseases and oldest known diseases and was first noticed by the farmers of Japan in 1884 and was called white withering disease. The seed-borne nature of the bacterium was first reported by Fang et al. (1956). The first Indian record of the disease was reported by Sreenivasan et al. (1959) from Maharashtra. In the 1960s, bacterial blight became prevalent in other rice-growing regions of Asia with the introduction of high-yielding cultivars line TNI and IR8, which were susceptible to the disease. The disease can cause severe yield loss of up to 50% depending on the rice variety, growth stage, geographic location, and environmental conditions. Losses due to the Kresek syndrome (seedling wilt) of BLB can reach as much as 75%. In India, losses in yield varied from 6%–60% in different states depending upon the stage and severity of infection and type of cultivars.

Symptoms

Bacterial blight is a vascular disease resulting in a systemic infection. There are three main symptoms caused by bacterial blight-leaf blight, wilt or Kresek, and yellow leaf or pale yellow.

(i) **Leaf blight:** Leaf blight symptom is most commonly seen and characterised by initial water-soaked, linear yellow to straw-coloured strips with wavy margins, mostly on both edges of the leaf (Figure 1.6A). These stripes usually start from the tip and extend downward. This is followed by drying and twisting of the leaf tip and rapid extension of marginal blight lengthwise and crosswise to cover the entire leaf blade. Lesions on older infected leaves often appear greyish to white. In severely diseased fields, brownish discolouration with slime can be seen on glumes (Figure 1.6B). In dry weather, opaque and turbid drops of bacterial ooze which dry into yellowish beads can be seen on the leaf surface which is splashed by rain (Figure 1.6C).

Figure 1.6 Bacterial Blight of Rice. (A) Straw colored strips with wavy margin on leaves, (B) Light to dark brown discoloration on glumes, (C) Yellow beads of bacterial ooze on infected leaf.

(ii) Wilt or Kresek: The wilt syndrome, known as *Kresek* (acute wilting of seedlings), is the most destructive form of the disease found at a temperature ranging from 28°C–34°C. It occurs in the tropics where younger plants less than 21 days old are the most susceptible. All leaves of infected plants wilt and roll up and turn greyish green. Eventually, these leaves turn yellow to straw-coloured and wither, and generally lead to the death of the whole plant. Plants that survive are stunted and yellowish. Total crop failure is also very common with Kresek phase.

(iii) Yellow leaf or pale yellow syndrome: Yellow leaf or pale yellow syndrome, less common, is associated with the bacterial blight of rice in the tropics. The youngest rice leaf becomes uniformly pale yellow or it has a broad chlorotic stripe. With yellow leaf, the bacteria are not present in the leaf itself but can be found in the internodes and crowns of affected stems.

Causal Organism and Etiology

Bacterial leaf blight (BLB) of rice is caused by *Xanthomonas oryzae* pv. *oryzae* (Xoo). The bacterium is a yellow, slime-producing, obligately aerobic, Gram-negative, non-spore-forming rod (0.55 × 3.5–2.17 μm) and monotrichously flagellated. Cells occur singly, in pairs, or sometimes in chains and filaments may also occur. Colonies on solid media containing glucose are round, convex, mucoid, and yellow due to the production of Xanthomonadin.

Disease Cycle and Epidemology

Xanthomonas oryzae pv. *oryzae* survives primarily in/on infected seeds, stubbles, straw, ratoons, self-sown plants, wild *Oryza* species (*O. rufipogon* and *O. australiensis*), rhizosphere of winter crops and in a number of weeds such as *Leersia oryzoides*, *Zizania latifolia*, *Leptochola chinensis*, *L. panacea*, *Cyperus rotundus*, *Cynodon dactylon*, *Cenchrus ciliaris*, *Brachiaria mutica*, *E. crusgalli*, *Panicum maxicum,* and *Paspalum scrobiculatum*. The primary infection may result from the inoculum overwintering in seed or in crop residue or on overlapping crops or soil and can infect nursery seedlings. The bacterium invades through wounds caused by root development or any other injuries occurred during handling, insect attack or natural openings like hydathodes and stomata on leaves. The common practice of cutting the leaf tips of seedlings before transplanting plays an important role in the development of kresek symptoms because the cut ends allow the entry of the bacterium directly into the xylem vessels. In addition, broken roots resulting from uprooting the seedlings from the seed-bed serve as entry points for bacteria present in flood-irrigated fields. Bacteria multiply in the intercellular spaces and when there is a sufficient bacterial population, some bacteria invade the vascular system and become systemic in the xylem of the rice plant. They enter veins and colonize the xylem vessels. Within a few days, the bacterium multiplies and produces EPS, fill the xylem vessel and bacteria ooze out from water pores (hydathodes), forming yellow beads or strands of exudate on the leaf surface and blade which acts as a source of secondary inoculum. EPS protects from the desiccation of the bacterial cells. The bacterium can be disseminated by irrigation water, by splashing or windblown rain, by plant-to-plant contact, by trimming tools used in transplanting, and by handling during transplanting. Bacteria that enter through roots, plug the xylem vessels and cause wilting. Infections and disease development are favoured by a temperature of 25°C–30°C, high humidity (above 70%), shading, heavy doses of nitrogenous fertilizers, rain, flooding, and severe winds.

Management

Cultural Methods

- Use the seed from a disease-free crop.
- Destry the diseased plant debris including rice straw and stubbles.
- Remove the weeds and collateral hosts.

- Nursery should be raised away from diseased fields and avoid flooding with contaminated water.
- Keep proper plant spacing.
- Do not apply excessive doses of nitrogen. Nitrogen should not be applied beyond six weeks after transplanting.
- Reduce nitrogen application and apply if needed only a small dose of nitrogen in more split. During vegetative growth and panicle initiation to the booting stage.
- Do not use pond water for irrigation.
- Do not grow the nursery under shade.
- Grow nurseries preferably in isolated upland conditions
- Avoid clipping seedlings during transplanting.
- Skip nitrogen application at booting (if the disease is moderate).
- Drain the field (except at the flowering stage of the crop).
- Avoid the flow of water from affected fields.
- Maintain proper plant spacing.

Physical Control

- Seed treatment with hot water at 57°C for 10 minutes or at 52°C–54°C for 30 minutes. This gives 95% disinfection of seeds.

Host Resistance

Deployment of gene-conferred host plant resistance provides an economical, effective, environment-friendly approach for managing plant diseases and minimising losses. Xa21 is the first resistance gene cloned in the monocots (rice). To date, more than 40 resistance (R) genes conferring host resistance to various strains of *Xoo* have been identified and 11 of them have been cloned, namely Xa1, Xa3/Xa26, Xa4, Xa5, Xa10, Xa13, Xa21, Xa23, Xa25, Xa27, and Xa41.

- Use of resistant varieties/cultivars having polygenic resistance, like Improved Pusa Basmati-1, NH-56, Jyothi, Ajaya, IR36, IR64, Swarna, Bhumbleshwari, PR111, PR 113, PR115, PR116, PR118, PR 121, PR 122, PR 123, PR 124, PR 127, Rajendra Basmati, Pant Dhan 10, Pant Dhan 11, Govind, Radha, Kamini, Jayshree, Kanchan, IR 20 IR 72, PONMANI, TKM 6, IR 20, Pant Dhan 19, Swarnadhan, Indra, Tholakari, Mahsuri, Tikkana, Pinakini, Deepti, Badava Mahsuri, MTU 9992, Godavari Ranjeet, Rajendran-201, Madhuri, Karjat-1, PR4141, BK79, ADT36, CO45, Sarjoo 52, Pant Dhan 4 and Improved Sambha Masouri.
- Use resistant variety/cultivars having polygenic resistance, like Improved Pusa Basmati-1, NH-56, Jyothi can also be used.
- Pusa 1460, a variety developed by pyramiding of resistance genes Xa5 + Xa13 + Xa21 in the background of Pusa Basmati-1 through marker-assisted backcross breeding has been released for cultivation in various states of India such as Punjab, Haryana, Jammu and Kashmir, and Delhi.

Biological Control

- Different strains of *Pseudomonas fluorescens* may be employed for disease management. Strains RRb-11 and MBPF-01 when applied as seed treatment and spray @ 10^8 CFU/ml respectively were demonstrated to be promising to reduce blight disease.
- Spray Neem oil @ 3% or NSKE 5%.
- Spray fresh cow dung extract to manage bacterial blight. Dissolve 20 g cow dung in one litre of water, then allow it to settle and sieve. Use supernatant liquid (starting from the initial appearance of the disease and another at fortnightly intervals).

Chemical Control

- Treat seed before sowing to kill primary inoculum.
- Treat the seeds with bleaching powder (100 g/L) and Zinc sulphate (2%) reduces bacterial blight.
- Treat the seed with Streptocycline (100 μg/L) + Agallol (0.1%) or Aretan/Tafasan (0.05%) followed by hot water treatment at 52°C–54°C for 30 minutes.
- In the nursery, spray Streptomycin sulphate (9%) + Tetracycline hydrochloride (1%) SP @ 100–150 ppm.
- Give two sprays of Nanocopper @ 0.2 ppm at fortnight intervals starting from active tillering stage.
- Spray Streptocycline @ 200 ppm at active growth phase.
- Spray Copper oxychloride @ 2.5% at fortnight interval starting from the active tillering stage.
- Spray Streptocycline 100–150 ppm solution at the early boot stage. Second spray, if necessary before grain set.
- Foliar sprays of Blitox-50WP (0.25%) + Streptocycline (100 μg/ml).
- Application of Bleaching powder @ 5 kg/ha in the irrigation water is recommended in the kresek stage.
- Foliar spray with Copper fungicides alternatively with Streptocycline (250 ppm) to check secondary spread.

1.7 TUNGRO DISEASE

Rice tungro, which means *degenerated growth* in a Philippine dialect. It is also called *mentek* or *habang* in Indonesia, *accepna pula* in the Philippines, *Penyakit merah* in Malaysia and *yellow-orange leaf* in Thailand. It is one of the most important viral diseases of rice prevailing in South and Southeast Asian countries and causes 30% to 100%. The annual loss global due to tungro is estimated approximately US $1.5 billion. In India, the occurrence of the disease was suggested by Raychaudhuri et al. in 1967 and confirmed by John in 1968. The outbreaks of rice tungro disease were reported in Bihar in 1969 and in Kerala in 1973. Rice tungro disease entered India with the import of and cultivation of the high yielding rice variety Taichun.

Symptoms

Rice plants infected with both *Rice tungro bacilliform virus* (RTBV) and *Rice tungro spherical virus* (RTSV), typically exhibit severe stunting, yellow or red or yellow-orange discolouration of infected leaves, reduced tillering, sterile panicles, and often irregular-shaped dark brown blotches on the leaves. The young infected leaves may have a yellow tip, stripping or chlorotic mottling, and interveinal chlorosis, while the older leaves show rust-coloured specks of varying size. Infected plants have delayed flowering. The panicles are small and not completely exerted, and bear mostly sterile or partially-filled grains often covered with dark brown specks. Plants infected with RTBV alone show similar but milder tungro symptoms. Plants infected with RTSV alone show no obvious symptoms except very mild stunting. In infected plants, RTBV is localised in the vascular bundles and RTSV in the phloem tissues. In infected cells, both RTBV and RTSV particles are scattered or aggregated in the cytoplasm. RTSV particles also occur in vacuoles. Viroplasm-like inclusions and membraneous masses occur in the cytoplasm of RTSV-infected cells.

Causal Organism and Etiology

Rice tungro is a composite disease, caused by the joint infection of two different viral species: *Rice tungro bacilliform virus* (RTBV) and *Rice tungro spherical virus* (RTSV). The RTBV is a pararetrovirus that belongs to the Genus: Tungrovirus, and Family: Caulimoviridae. The virus particle is bacilliform, 100–400 nm long and 30–35 nm in diameter, circular dsDNA genome. RTBV alone causes severe yellowing of leaves and mild stunting symptoms in the host but is incapable of independent transmission through green leafhoppers. RTSV is a plant picornavirus, belongs to the Genus: Waikavirus and Family: Secoviridae. The virus particle is isometric with diameter of 30–33 nm, monopartite, positive-sense-ssRNA genome. RTSV can be transmitted independently by green leafhoppers, causes only very mild stunting symptoms in most rice varieties but it intensifies the symptoms of tungro caused by RTBV.

Disease Cycle and Epidemiology

Tungro viruses are transmitted by green leafhoppers. The main tungro vector, *N. virescens*, is monophagous and is restricted to *O. sativa* and some closely related wild rices. Transmission mainly by the leaf hopper vector *Nephotettix virescens* males, females, and nymphs of the insect can transmit the disease. Both the virus particles are jointly transmitted semi-persistently, in the vector the particles are noncirculative and nonpropagative. Symptoms appear in 21–28 days old plants. Plants infected with RTSV alone may be symptomless or exhibit only mild stunting. RTBV enhances the symptoms caused by RTSV. RTSV can be acquired from the infected plant independently of RTBV, but acquisition of RTBV is dependent on RTSV which acts as a helper virus. The transmission rate is high at 34°C and efficiency of females is better than males of leaf hopper. Minimum acquisition feeding period is 5–7 min but transmission increases with acquisition access feeding upto 4 days. There is no latency. Vector becomes viruliferous immediately after feeding. The viruses are retained by the vector for up to 5 days. Weed hosts play an important role in survival and dissemination of rice tungro virus including *Eluesine indica, Echinochloa crusgalli, Leersia hexandra, E. colona, Echinochloa glabrescens*, and *Leptochloa chinensis*. Both the viruses thrive in rice and several weed hosts which serve as source of inoculum for the next. These viruses do not survive in seed or soil. Ratoon from infected rice stubble serves as reservoirs of the virus. Disease incidence depends on rice cultivars, time of planting, time of infection and presence of vectors, and favourable weather conditions.

Management

Cultural Methods

- Adjust the date of planting.
- A fallow period of at least a month is recommended to eliminate hosts, viruses, and vectors of the disease.
- In epidemic areas follow rotation with pulses or oil seeds.
- Apply neem cake @ 12.5 kg per 20% nursery as basal dose.
- Plouging and harrowing the field to destroy stubbles right after harvest to eradicate other tungro hosts are also advisable.
- Remove the weed hosts from the bunds.
- Apply neem cake to nursery ploughing to incorporate stubbles.
- Set up light traps to attract and manage the leaf hopper vectors as well as to monitor the population.

Host Resistance

- Planting resistant varieties against tungro virus disease is the most economical means of managing the disease.
- Use Resistant varieties like IR 36, IR 50, ADT 37, Ponmani, Co 45, Co 48, Surekha, Vikramarya, Bharani, IR 36 and white ponni.

Chemical Methods

- Spray 2% urea mixed with Mancozeb @ 2.5 g/L to minimise the leaf yellowing.
- Instead of urea foliar fertilizer like multi-K (Potassium nitrate) can be sprayed @ 1% which imparts resistance.
- Spray two rounds of Monocrotophos 36 WSC (40 ml/ha) or Fenthion 100 EC (40 ml/ha) at 15 and 30 days after transplanting. The vegetation on the bunds should also be sprayed with insecticides.
- Apply Carbofuran granules @ 1 kg/ha or Quinalphos 5 G @ 2.0 kg/ha in the nursery when virus infection is low to manage vector population.
- Apply Carbofuran granules @ 3.5 kg/ha or spray Monocrotophos @ 1.6 to 2.2 ml/L to manage insect vector during pre-tillering to mid-tillering if one affected hill/m is observed.

1.8 KHAIRA DISEASE

Zinc deficiency was first diagnosed in rice on calcareous soils of Northern India by Y. L. Nene, in 1966. Zinc (Zn) deficiency is now considered the most widespread nutrient disorder in lowland rice.

Symptoms

Zinc deficiency causes multiple symptoms that usually appear 2–4 weeks after rice transplanting. The symptoms include dusty brown spots on the upper leaves of stunted plants, patches of poorly established plants, chlorotic brown midrib near the leaf base of younger leaves, brown blotches and streaks on lower leaves, increased spikelet sterility, reduced leaf blade size, and sometimes white line appears along the leaf midrib, delayed maturity and poor yields. In severe cases, stunted plants may die, while those that recover show the substantial delay in maturity and reduction in yield.

Causal Agent

Khaira disease is caused by Zinc deficiency. Zn deficiency affects several biochemical processes in the rice plant, thus severely affecting plant growth. Zn deficiency leaf symptoms resemble Sulphur (S) and Iron (Fe) deficiency in alkaline soils and iron toxicity in poorly drained organic soils. Zinc is an essential plant nutrient required for several biochemical processes in the rice plant, including chlorophyll production and membrane integrity. Thus, Zn deficiencies affect plant colour and turgor. Zn is only slightly mobile in the plant and quite immobile in the soil.

Occurrence of Zn Deficiency

Zn deficiency is associated with a wide range of soil conditions: intensively high soil pH (above 7.0), low available Zn, prolonged submergence and low redox potential, high organic matter and bicarbonate content, high magnesium (Mg) to calcium (Ca) ratio, and high available phosphorus, cropped soils; continuously flooded paddy soils or very poorly drained soils.

Management

- Grow Zn-efficient varieties.
- Rice plants can recover from Zn deficiency if the field is drained—a dry fallow increases the availability of Zn.
- Apply the Zn-fertilizer on the soil surface after the last puddling and leveling in the main field or apply Zn-fertilizer to the nursery beds 7–8 days before uprooting seedlings.
- Broadcast $ZnSO_4$ in a nursery seedbed.
- Dip seedlings or presoak seeds in 2%–4% Zinc oxide (ZnO) suspension (e.g., 20–40 g ZnO/L of water).
- It is enough to apply Zinc sulphate @12.5 kg/ha, if green manure (6.25 t/ha) or enriched FYM, is applied.
- Apply 25 kg of Zinc sulphate with 50 kg sand before transplanting.
- If Zn deficiency symptoms are observed in the field, apply 10–25 kg Zinc sulphate ($ZnSO_4 \cdot H_2O$) or 20–40 kg Zinc Sulphate Heptahydrate ($ZnSO_4 \cdot 7H_2O$) per ha on the soil surface suspension before planting.
- The crop only requires around 0.05 kg Zn/ha (both straw and grain) per ton of grain yield, but much more Zn fertilizer must be applied because Zn once applied is not very available to the plant.

- Apply Zinc sulphate @ 5–10 kg/ha or apply it as a foliar spray @ 0.5%–1.5% at tillering (25–30 days after tillering), repeat 2–3 applications at 10–14 days intervals.
- Zinc chelates (e.g., Zn-EDTA) can be used for foliar application.
- Use fertilizers that generate acidity (e.g., replace some urea with Ammonium sulphate).
- Apply organic manure before seeding or transplanting or to the nursery seedbed a few days before transplanting.
- Allow permanently flooded fields to drain and dry out periodically.
- The effect of Zn application to soil can last 2–5 crop seasons on all soils except in alkaline soils. In alkaline soils, Zn may need to be applied to each crop.

Maize (*Zea mays* L.)

2.1 STALK ROTS

Stalk rot is one of the most important diseases of corn worldwide. The term *stalk rot* is widely used to connote the premature senescence and deterioration of basal stalk tissues near the physiological maturity of corn plants. It is characterised by premature plant death and soft, often broken, lower stalk internodes. It can be caused by many fungi and bacteria, which primarily infect senescing, injured, or stressed plants near maturity. Yield losses caused by stalk rot depend on the aggressiveness of the pathogens and the development stage of the plant. The yield losses can range from 12%–40% by stalk rot in maize. Losses are caused either by poor filling of the cobs or due to the lodging of affected plants. Some pathogens, such as *Fusarium* spp., *Colletotrichum graminicola*, and *Macrophomina phaseolina* can colonize the stalk tissues of apparently healthy plants during the vegetative stage, causing reduced grain weight and severe yield losses. The main cause of yield losses due to stalk rot is the colonization and degradation of stalk vascular tissues by pathogens, hindering the translocation of water and nutrients to the plants. Therefore, the premature death of plants, stalk breakage, and lodging are common in areas with a severe occurrence of the disease. The following pathogens are associated with the stalk rot of maize. The difference between diffrent stalk rot of maize is given in Table 2.1.

2.1.1 Anthracnose Stalk Rot

Symptoms

Anthracnose diseases of corn include stalk rot, top dieback, and foliar and seedling diseases. The most visible symptom of the stalk rot phase is shiny, black blotches on the lower internodes of the stalk. When pinched, the stalk collapses easily and one can see that the discolouration extends into the pith. The fungus produces asexual fruiting structures, called acervuli that contain tiny, dark, whisker-like appendages, called setae, on infected tissue. In the top dieback phase, the upper part of the plant appears bleached, while the lower plant tissue remains green. Black lesions may be present on the stalks in the top internodes when the leaf sheaths are removed. Lodging and breaking of stalk occur and breaking is higher lengthwise than to other stalk rots. Sometimes, stalk rot symptoms first occur in the upper canopy, it is referred to as anthracnose top dieback which is a phase of the stalk rot disease.

Table 2.1 Difference and management of different of stalk rots of maize

Disease and causal organism	Signs and symptoms	Favorable weather for infection	Management strategies
Anthracnose (*C. graminicola*)	Stalk: Black, shiny lesions, often blotchy in appearance. Pith: Disintegrated.	Cloudy days with high humidity, temperature 25°C–30°C.	• Plant hybrids with some resistance. • Reduce plant stress and wounding. • Rotate crops (avoid rotating to sorghum).
Charcoal rot (*M. phaseolina*)	Pith: Disintegrated and covered in silver-black fungal structures ("dust"). Stalk and roots: May be covered in black fungal structures.	Hot and dry conditions (high soil temperatures 30°C–42°C), especially near plant maturity.	• Irrigate during hot, dry seasons. • Avoid common plant stressors (heavy foliar disease, nutrient imbalances, etc.).
Gibberella stalk rot (*G. zeae*)	Pith: Disintegrated with pink-red discolouration. Stalk: Small, black fungal structures (perithecia) (easily scraped off).	Warm (25°C–28°C), moist conditions.	• Avoid wheat-corn rotations. • Avoid plant stress and wounding.
Diplodia stalk rot (*S. maydis*)	Pith: Disintegrated. Stalks: Brown/black fungal structures embedded in the rind (not easily scraped off).	Warm (26°C–31°C), moist conditions.	• Rotate crops. • Plant hybrids with some resistance. • Till infected residue into the soil. • Avoid plant stress and wounding.
Fusarium stalk rot (*F. verticilliodes, F. proliferatum, and F. subglutinans*)	Pith: Disintegrated with salmon-pink discolouration. Leaves: Dull green.	Dry conditions before silking followed by warm (26°C–38°C), and wet conditions after silking.	• Reduce plant stresses. • Control foliar diseases. • Provide adequate field drainage. • Avoid high-nitrogen, low-potassium soils.
Pythium stalk rot (*P. aphanidermatum*)	Pith and rind: Soft, brown, and water-soaked. Leaves: Green even after lodging. Stalk: Twists during lodging.	Hot (25°C–35°C), wet, and humid conditions.	• Avoid overhead irrigation using a pond or other stagnant water. • Provide adequate field drainage. • Avoid common plant stressors.
Bacterial stalk rot (*E. chrysanthemi* pv. *zeae*)	Stalk and rind: Dark-brown, water-soaked lesions; soft, slimy tissue; foul odor; twists during lodging. Plant tips: Slimy and rotten; may die prematurely.	Hot (temperature range 28°C–30°C) and humid periods.	• Avoid overhead irrigation using a pond or surfacewater. • Provide adequate field drainage. • Avoid common plant stressors, including insect damage.

Causal Organism and Etiology

Anthracnose stalk rot of maize is caused by *Colletotrichum graminicola*. The aerial mycelium is dark grey with orange spore masses. Conidia produced in acervuli are falcate, slightly curved and tapered toward the tips. The aerial mycelium of the pathogen is dark grey with orange-coloured spore masses, and conidia are falcate, slightly curved, tapering toward the tips, and produced in acervuli with setae.

Disease Cycle and Epidemiology

The corn anthracnose cycle can be characterised by five temporal phases: production and dissemination of primary inoculum, seedling blight by primary inoculum, leaf blight by repeating secondary inoculum, systemic colonization, and stalk rot, and saprophytic survival. The fungus overwinters in corn residue. Conidia formed on leaves may be washed behind the leaf sheath and initiate rind infection. The fungus can infect roots, or rain and wind can disperse fungal spores from plant residues to corn stalks. Infection occurs primarily through wounds caused by insect feeding, hail, mechanical damage, or other mechanisms. Seedlings can be infected, and some plants may die before pollination. Anthracnose can infect corn at any point in the growing season, but infection is favoured by cloudy, warm, and humid weather after silking. The fungus grows biotrophically for a relatively short time (24–36 hours), after which it penetrates host cells and grows necrotrophically. The epidermal cells of non-injured, living stem rind tissues appear to be invaded in a biotrophic manner from appressoria similar to infection of the foliar epidermis. Pathogen infection of internal tissues of living stems occurs primarily through wounds that breach the stalk rind. Conidia germinate at wound sites, producing extremely long germ tubes. Corn cells are usually penetrated from appressoria formed where the germ tubes come in contact with cell walls, but occasionally, penetration at wound sites occurs from hyphal strands without the formation of appressoria. The pathogen rapidly becomes established as a necrotroph and causes significant maceration of the host tissue. Once the xylem vessels are invaded, the fungus grows systemically in the stalk. However, anthracnose development is favoured by extended periods of high humidity, a condition necessary for the sporulation of the fungus. The disease is also favoured by rain since conidia are most easily dispersed by splashing raindrops. Although germination and appressorium formation by *C. graminicola* occur over a broad temperature range (15°C–35°C), penetration of the host appears to occur only in the narrow range of 25°C–30°C. The presence of melanin in the wall of the appressorium is believed to allow for the development of exceptionally high hydrostatic pressures within the appressorium, which facilitates penetration of the host cell wall. The infection process is aided by the presence and/or production of various enzymes by the fungus, including cutinases, cellulase, pectinases, and polygalacturonases, which account for the degradation of the cutin component of the plant cuticle and host cell wall materials, respectively. As with many other *Colletotrichum* species, *C. graminicola*, upon penetration of the host cell, appears to grow first as a biotroph. Once established in the leaf tissue, the pathogen begins to parasitize the host as a necrotroph, producing a variety of cell wall–degrading enzymes and killing the host tissue in advance of the now intracellularly developing mycelium. Conidia germinate at wound sites, producing extremely long germ tubes. Corn cells are usually penetrated from appressoria formed where the germ tubes come in contact with cell walls, but occasionally, penetration at wound sites occurs from hyphal strands without the formation of appressoria. The pathogen rapidly becomes established as a necrotroph and causes significant maceration of the host tissue. Once the xylem vessels are invaded, the fungus grows systemically in the stalk.

2.1.2 Charcoal Rot

Symptoms

After flowering, initial symptoms are the abnormal drying of upper leaf tissue, stem lodging, and premature death. At maturity, the lower stem internodes (usually limited to the first 5 nodes) show a typical charcoal, grey-black discolouration. When a stem is cut open, numerous minute black specks (microsclerotia) are visible on the shredded vascular bundles and on the inside of the stem, giving the interior parts of the stem a charred appearance. Inside the stalks numerous tiny microsclerotia which are long-term, thick-walled survival structures are present that give a speckled appearance, similar to a silvery black charcoal dust. Brown, water-soaked lesions, which later turn black, are present on the roots. Microsclerotia may also be visible on the roots and just below the stalk surface. Infected kernels show pale yellow with black streaking below the pericarp, chaffy and loose ear. Kernels are easily removed from the cob, and they show small, round, black, pinhead-like sclerotia on the surface.

Causal Organism and Etiology

Charcoal rot of maize is caused by *Macrophomina phaseolina*. The fungus can also infect alfalfa, sorghum, and a variety of weed species. Mycelium is dark grey-green. Microsclerotia are small, round, black, and pinhead. Pycnidium is black and globose with ostiolate apically. Conidiophores are hyaline, simple, cylindrical, and narrowing apically. Conidia are hyaline, cylindrical, and single-celled.

Disease Cycle and Epidemiology

Charcoal rot is a soilborne disease. *M. phaseolina* overwinters as sclerotia in crop residues and soil and can remain viable for several years. The alternate hosts are also a major source of inoculum. In dry and hot conditions fungi infect the roots of maize plants and colonize the lower stalk, eventually giving rise to characteristic symptoms (abundant, black micro-sclerotia and charring and shredding of the pith tissue). The pathogen disperses through the movement of soil, contaminated with microsclerotia, on tractors, ploughs and other farm machinery and packing material. The severity of disease caused by *M. phaseolina* in various hosts is associated with high soil temperatures (30°C–42°C) and low moisture or when plants are under abiotic stresses. This disease is typically a problem in extremely hot, dry seasons when there is low soil moisture, particularly during the grain-filling period. Maximum infection in plants occurs under moisture stress during the post-flowering period. Post-flowering stresses due to high plant population or drought coupled with heavy applications of nitrogen fertilizer, hail or insect damage promote disease development.

2.1.3 Gibberella Stalk Rot

Symptoms

Usually, infection closely follows pollination. Signs to look for when scouting for Gibberella stalk rot are tiny, round, blue-black fungal structures (called perithecia) on the surface of the stalk near the first internode. One can easily remove the perithecia with a fingernail. Inside the stalk, the pith will be disintegrated, leaving only the stringy vascular bundles. The pith also will have a pink-red colouration. Infection can occur through roots, stalk sheaths, or leaf sheaths and progress to infect the stalk.

Causal Organism and Etiology

Fusarium graminearum (teleomorph: *Gibberella zeae*) is the causal agent of Gibberella ear rot, seedling blight, root and stalk rot of maize depending on the timing and tissue of infection. The asexual stage of the fungus produces macroconidia, and the sexual stage produces spores called ascospores.

Disease Cycle and Epidemiology

Fusarium graminearum overwinters on infested crop residues (corn stalks, wheat straw, and other host plants). On infected crop residues, the fungus produces macroconidia (asexual spores) which are desiminated by rain-splash or wind to plants and other plant debris. In the new croping season, when conditions are warm, humid, and wet, the sexual stage (bluish-black sexual fruiting bodies, perithecia) of the fungus (*Gibberella zeae*) develops on the infected plant debris. The sexual spores (ascospores) forcibly discharge from these sexual fruiting bodies (perithecia) into the air. The ascospores are picked up by turbulent wind currents and may travel great distances in the air. Infection may be caused by ascospores or macroconidia through the silks (ears), leaf sheath or roots. The pathogen establishes itself inside the plant before it starts killing host tissues. Warm weather (25°C–28°C) with high humidity favours the disease development. Abiotic stresses such as water logging aggravate the disease. The disease is most prevalent 21 days after silking.

2.1.4 Diplodia Stalk Rot

Symptoms

Plant leaves may turn a dull green as the fungus destroys the pith. The pith disintegrates, leaving only the vascular bundles, weakening the stalk and predisposing plants to lodging during strong winds and rain. No red or pink discolouration of the pith, as is noticed with *Gibberella* and *Fusarium* stalk rots. Internally, mats of white fungal growth may be evident on affected tissues. The pathogen also produces diagnostic signs on the stalk's surface as small, black, flask-shaped, spore-producing structures called pycnidia. Later in the season, the most conspicuous symptom is the abundant presence of minute, dark brown to black specks (pycnidia) embedded in the rind tissue near the damaged internodes where rotting has occurred. The pycnidia can give the stalk a sandpaper-like texture. Unlike the blue-black fruiting bodies (perithecia) of Gibberella stalk rot can easily be scraped off the stalk's surface, the pycnidia of Diplodia stalk rot are embedded in the stalk tissue and cannot be easily removed by scraping the tissues.

Causal Organism and Etiology

Diplodia stalk rot, caused by the fungus *Stenocarpella maydis* (previously known as *Diplodia maydis*), has become one of the most important stalk rots in recent years, and it is the same fungus that causes Diplodia ear rot.

Disease Cycle and Epidemiology

S. maydis survives as pycnidia in stalk debris that is either buried or on the soil surface in between growing seasons. The fungus is also seedborne. Under warm, moist conditions, spores are released from fruiting bodies and spread by rain and wind. The fungus primarily invades through the lower stalk tissue, crown, and roots. Infection is favoured by dry weather before silking, followed by warm, wet weather after silking. Corn is most susceptible during the post-silking period. Ear infection occurs is most common during rain when. The ear rot phase is favoured by dry, followed by warm (26°C–31°C), and wet conditions, approximately two weeks before and after silking. The disease complex is most severe under conditions of continuous cropping, and reduced or no-till practices. Other factors that can favour disease development include: too high or low fertility, high plant populations, and injury from insects, disease, or severe weather.

2.1.5 Fusarium Stalk Rot

The pathogens are seed borne with up to 75% of all seeds planted infected with *F. verticillioides*. The stalk rots are present worldwide where maize or cereals are cultivated. Plants are more susceptible to other stresses such as other pathogens, drought, or other abiotic factors. Infected plants fall down which may cause high yield losses.

Symptoms

Infected plants typically wilt, leaves turn dull greyish-green and the lower stalk turns from dark green to straw-coloured. Rotting occurs at roots, crown, and lower internodes. When split opened, inner stalk shows a light pink to tan discolouration. Pith disintegrates but vascular bundles remain intact.

Causal Organism and Etiology

Fusarium ear rot of maize is caused by *Fusarium verticillioides.* It also infects a range of cultivated crops, including sorghum, sugarcane, wheat, cotton, banana, pineapple, and tomato. Macroconidia are hyaline, slightly curved, tapered at both ends, and 3–5 septate. Microconidia are single-celled, produced prolifically, and are borne in chains. Asexual spores develop outside, at the nodes in rings. Sexual bodies and spores develop on the basal part of the stalk. The genetic material of these fungi varies a lot by sexual multiplication and anastomosis between mycelia and asexual spores.

Disease Cycle and Epidemiology

The pathogen *Fusarium verticillioides* overwinter in the soil and the infected plant residues as mycelia, thick-wall chlamydospores, or sexual bodies in some species and dispersed by wind and rain splash. Corn borer adults have been shown to vector the disease from plant to plant. Corn borer larvae create wounds that allow the fungus to enter the plant. The maize stalks are attacked by sexual or asexual spores of the fungus. Loose tissues and injuries are the best entry point for infection. The pathogen can infect the plants directly through the roots, causing root and lower stalk rot. It can also infect the nodes when dispersed to leaves and washed down into the sheath. The pathogen colonizes the internal parts of the internodes and the stem remains empty. The mycelia develop chlamydospores when the nutrients are exhausted in the plant tissue. The disease is favoured by warm, relatively dry weather; plant stress following pollination; and other diseases. The disease generally progresses during the reproductive stages of corn development. The disease typically occurs in a complex with other root or stalk rots including *M. phaseolina*.

2.1.6 Pythium Stalk Rot

Pythium stalk rot is generally confined to hot and humid regions where maize is cultivated in poorly drained soils.

Symptoms

Infection is typically confined to the first internode directly above the soil surface. Stalks become soft, water-soaked, darkened, and collapsed. Like bacterial stalk rot, lodging occurs in a twisting fashion instead of cracking. Lodged corn plants may appear healthy for extended periods as vascular bundles often remain intact for some time. Pythium stalk rot can occur any time during the growing season but is most frequently prior to flowering or tasseling stage.

Causal Organism and Etiology

Pythium stalk rot of maize is caused by *Pythium aphanidermatum*, an Oomycetes. Hyphae are non-septate and hyaline. Sporangia are inflated (balloon like), branched or unbranched. Zoospores are kidney-shaped, laterally biciliate. Oogonium is spherical with straight stalks. Antheridium is intercallary. Oospores are thick-walled and typically oospores can be observed in diseased stalk tissue.

Disease Cycle and Epidemiology

Pythium aphanidermatum survives as oospores, sporangia, or mycelia in infested soil residue. Oospores can survive in the soil for several years. During humid weather at temperatures of 32°C and higher or where air and soil drainage is poor, sporangia germinate either directly to form a germ tube or indirectly by releasing zoospores into the soil. In presence of free water, zoospores swim toward plant or root tissue in response to exudates and cause infection directly at or just below the soil surface, leading to the development of characteristic lesions. In the sexual stage of the life cycle, the pathogen produces oospores within the infected host tissues.

These oospores are able to overwinter in crop debris or the soil. The pathogen is dispersed with either the movement of infected crop debris or with flooding or excess wetness, which transports oospores and enables zoospores to swim freely. Pythium stalk rot is more prevalent and infection is favoured by extended periods of high temperatures (25°C–35°C), wet, high humid weather, high crop density, and high levels of nitrogen fertilizer, and disease occurrence is worse in fields with poor soil drainage.

2.1.7 Bacterial Stalk Rot

Bacterial stalk rot is a major disease of maize in tropical and subtropical countries. It is particularly severe under conditions of high temperature and humidity. It is one of the four major stalk diseases of maize in India. Incidence up to 80%–85% has been observed in nature and yield losses of 98.8% in artificial epiphytotic.

Symptoms

Symptoms appear late on 40–60 days old maize plants. The typical symptoms are soft rot and darkening of the tissues affecting the stalk and the top of the plant, causing the breaking of the stalk. The lesions progressed from the top to below nodes, leaf sheaths, and blades and rotten tissues emitted an unpleasant odor.

Top rot: The first visible symptoms are withering or wilting and drying up of the tips of the middle leaves of the whorl. At the same time, a soft rot develops in the stalk at the base of the whorl. Leaves become yellow and the infected tissue becomes brown, soft, and water-soaked. A rapid decay downwards through the stalk is observed and the tops of affected plants droop. The cluster of leaves can easily be pulled out, showing a soft, rotted condition at the breaking point near the base of the whorl. When the ears along with the husks are infected, they first become water-soaked, turning slimy and, later, drying up. The plants may bear undeveloped ears with many rotting grains and a covering of slime, or the infected ear completely rots (cob rot) and does not bear any seed.

Basal rot: Internally, the stalk turns into a soft mass of disintegrated tissue. At this stage, the plants usually topple over. A foul odor and the presence of dipterous larvae on and in decaying tissues are characteristic symptoms of this disease and help to distinguish it from the fungal stalk rots. The rotting stalk material often appears dark brown, mushy, and slimy. Lodging often occurs in a twisting, rather than cracking, manner. The rot may involve only one or two internodes, or the entire length of the stalk, which finally dries up and its interior turns into a shredded mass of fibrous tissue. Since the bacteria do not degrade the vascular tissue, the plant will likely remain bright green.

Causal Organism and Etiology

Bacterial stalk rot of maize is caused by *Dickeya zeae*. The cells are Gram-negative rods (0.6–0.8 × 1.5–3.0 μm) with rounded ends, motile by peritrichous flagella, facultative anaerobes, and metabolism is fermentative.

Disease Cycle and Epidemiology

The pathogen is soilborne and may survive in crop residues in the soil. It attacks the plant through its natural openings like the stomata, hydathodes, wounds in the leaf whorl, stalks, and roots caused by insects; and injuries brought about by strong winds or mechanical means. Plant debris has important role in the long survival of *D. zeae*. The bacteria can survive almost nine months in soil containing naturally infected maize stalks. High temperature and soil moisture promote the build-up of bacterial population and its continuous survival. It can also be spread from plant to plant and field to field through rainwater and its runoff. Certain insects like maize borer play a vital role in the initial infection and subsequent spread of the disease. The optimum temperature for disease development is 28°C–30°C and relative humidity above 90% also favours the disease development. Heavy doses of nitrogenous fertilizers increase the incidence of the disease.

Management Fungal Stalk Rots

- Use of resistant varieties
- Low plant density.
- Apply the balance dose of fertilizers.
- Manage insect pests (such as corn borer and fall armyworm to prevent plant wounds and stress).
- Do timely harvesting of corn.
- Test soils regularly and apply nutrients based on soil test results and yield goals. Be sure potassium levels are adequate, and manage nitrogen to prevent losses and ensure its availability throughout plant uptake.
- Do management of crop residue. Stalk rot pathogens overwinter in corn residue. The occurrence and intensity of stalk rots are sometimes related to the amount of inoculum present.
- *F. graminearum* stalk rot is managed by reducing inoculum through crop rotation and controlling insect wounding that could facilitate pathogen entry and infection.
- Scouting of corn field: Careful scouting and harvesting fields according to crop conditions can help prevent field losses due to stalk rot.
- Treat the seeds with Thiram @ 2.5 g/kg of seeds.
- Spray the crop with Carbendazim @ 0.25% for fungal stalk rot.
- Apply Captan in poorly drained pockets of the field along the rows at the rate of 150 g per 100 L of water.

Management of Bacterial Stalk Rot

- Remove and destroy diseased plant debris.
- Avoid waterlogging of the field.

- The number of plants should not exceed the recommended population density (55,000/ha).
- Field should have proper drainage.
- Planting of the crop on ridges rather than flat soil.
- Avoid excessive irrigation using surface water (such as sewage, ponds, lakes, or slow-moving streams) that can potentially harbor the pathogen.
- Select the well drained field or arrange proper drainage to avoid water logging.
- Avoid injury on the plants at the time of weeding and top dressing of urea.

Host Resistance

- Use of resistant cultivars such as PAU 352, PEMH 5, DKI 9202, DKI 9304.

Chemical Control

- Soil drenching with chlorinated water (100 µg m/L Chlorine) at 2-week intervals, starting from knee-high stage to the flowering stage, reduced the disease incidence by 75.92%.
- Two applications of bleaching powder @ 25 kg/ha. The first application should be given at the flowering and the second, 10 days after the first application.
- Soil drenching of bleaching powder containing 33% chlorine @ 10 kg/ha as at pre-flowering stage.
- Bleaching powder should be applied along the rows @ 10–25 kg/ha.
- Spray with Copper oxychloride @ 1% or Sterptocycline @ 100 ppm/ha.

2.2 DOWNY MILDEWS

The first record of downy mildew on maize in India was made by Butler (1913). Downy mildew of maize is one of the most destructive diseases in tropical regions of the world, particularly in many regions of tropical Asia, and can cause losses of upto 70%. The maize crop is uniquely attacked by ten different downy mildew pathogens belonging to the genera *Peronosclerospora, Sclerophthora*, and *Sclerospora*, around the world. These are sorghum downy mildew of maize (*Peronosclerospora sorghi*), Philippine downy mildew (*Peronosclerospora philippinensis*), sugarcane downy mildew (*P. sacchari*), brown stripe downy mildew (*Sclerophthora rayssiae* var. *zeae*), graminicola downy mildew or green ear (*Sclerospora graminicola*), crazy top (*Sclerophthora macrospora*), spontaneum downy mildew (*P. spontaneae*), Rajasthan downy mildew (*P. heteropogoni*), and leaf splitting downy mildew (*P. miscanthi*). These pathogens are an obligate parasite that belongs to Oomycetes. The symptoms produced by most of these species are more or less similar on maize leaves (Table 2.2, Figure 2.1). Infection by maize streak virus masks the symptoms of downy mildew. The most common and destructive downy mildews occurring in India are described below:

Table 2.2 Difference in downy mildews of maize

Downy mildews	Symptoms	Brief disease cycle and epidemiology
Brown stripe downy mildew of maize (*Sclerophthora rayssiae* var. *zeae*)	Narrow chlorotic spots with well-defined margins that extend in a parallel fashion between the veins appear on the leaf plates.	Soilborne
Sorghum downy mildew (*Peronosclerospora sorghi*)	Development of long, rather broad, chlorotic stripes on the leaf which are initially yellow and finally turn brown.	The pathogen can perennate through oospores in soil and on alternate host.
Philippine downy mildew of maize (*Sclerospora philippinensis* or *Peronosclerospora philippinensis*)	Leaves become pale yellow with wooly white growth on the under surface, plants possess shorter internodes, and infected plants may not produce any cob.	Pathogen survives on the alternate host.
Sugarcane downy mildew (*Peronosclerospora sacchari* or *Sclerospora sacchari*)	Main symptom is the development of long rather broad chlorotic stripe along the entire length of leaf, young leaves show downy growth more clearly than older leaves.	Pathogen can perennate through oospores in soil and on the alternate host.
Rajasthan downy mildew (*Peronosclerospora heteropogoni*)	Typical symptom is "*half diseased leaf*", infected leaves show yellow stripes at the base that also extends up to the upper green portion.	Soilborne
Crazy top (*Sclerophthora macrospora*)	The affected plants show "leafy" tassel, excessive tillering, severe stunting, narrow (strap-like) leathery leaves and long yellow to brown stripping on the leaves.	Pathogen is seed and soilborne.

Figure 2.1 Downy Mildew of Maize.

2.2.1 Brown Stripe Downy Mildew (BSDM)

Symptoms

Sclerophthora rayssiae var. *zeae* causes leaf lesions only. In the early stages of infections, leaves show narrow chlorotic or yellowish stripes, 3 to 7 mm wide. In some maize genotypes, these stripes may be reddish to purple. The lesions have well-defined margins and extend parallel with and are delimited by the leaf veins. Advanced striping and blotching occur with the confluence of adjacent lesions. The disease may first be noticed on the lower leaves, which will show the greatest degree of striping. The pathogen does not infect systemically to the plant.

Causal Organism and Etiology

Brown stripe downy mildew of maize is caused by *Sclerophthora rayssiae* var. *zeae*. Sporangiophore: short, determinate, and produced from hyphae in the substomatal cavities. Sporangia: formed sympodially in groups of 2–6, arising in basipetal succession. Sporangia are hyaline; ovate, obclavate, elliptic or cylindrical; smooth-walled; and are papillate, possessing a projecting truncate, rounded or tapering poroid apex. The sporangia are caducous, with a persistent, straight or cuneate peduncle. Four to eight zoospores are formed in the sporangia and may encyst within or outside of it. Zoospores are hyaline and spherical. Oogonia are hyaline to light. Thin-walled, they may have one or two paragynous antheridia. Oospores are pleurotic, spherical, or subspherical; and are hyaline, with one prominent oil globule. Cell walls are smooth, glistening, and confluent with the oogonial wall.

Disease Cycle and Epidemiology

Primary inoculum comes from oospores overseasoning in soil or plant debris or from mycelium in infected seed. The seed surface may carry plant debris containing viable oospores and the seed may carry oospores or mycelium within the embryo. The weed host *Digitria sanguinalis* also plays an important role in the survival and spread of *Sclerophthora rayssiae* var. *zeae*. Oospores generally undergo indirect germination, producing sporangiophores that bear sporangia which may contain four to eight zoospores. Less frequently, the sporangium may germinate directly and produce a germ tube capable of penetrating maize leaves. The rapid spread of the pathogen in the field occurs with the production of sporangia (secondary inoculum), which are dispersed in wind and water splash, or from physical contact with an infected plant. Sporangia production, germination, and infection require a film of water. Leaf wetness of 12 hours is required for infection via zoospores, with longer periods producing greater numbers of infected plants. Young plants are most susceptible but the susceptibility decreases with plant age. The optimum temperature for sporulation is 22°C–25°C and for zoospore production is 20°C–22°C. High temperature (28°C–33°C) and high relative humidity (100%) are conducive to disease development.

2.2.2 Sorghum Downy Mildew

Symptoms

Systemic infection occurs only in young plants. Initially, chlorotic to white streaks develop parallel to the leaf midrib. Sometimes half become chlorotic. A downy growth consisting of conidia and conidiophores appears in chlorotic tissue on both surfaces of the diseased leaf. Severely diseased plants are chlorotic, striped, stunted, sterile, and have narrow, erect leaves. Tassels are phyllodied, with floral parts converted to small leaves, which results in an abnormal seed set.

Causal Organism and Etiology

Sorghum downy mildew of maize is caused by *Peronosclerospora sorghi*. It is an obligate parasite. The fungus produces coenocytic mycelium. The sterigmata are longer than those of *S. graminicola*. The major difference lies with the sporangia which always germinate by germ tube (conidia-like), not produce zoospores. The sporangium is commonly spherical instead of oval and lacks the apical papilla. Oospores is spherical and three walls, the exosporium, mesosporium and endosporium. The endosporium is smooth, yellow in colour and of even thickness.

Disease Cycle and Epidemiology

The major inoculum source for *P. sorghi* is oospores which can survive for several seasons in soil. Susceptible seedlings are attacked through underground parts, which become systemically infected. Airborne conidia, which are produced on the leaf surfaces, serve as secondary inoculum that can induce systemic infection in plants up to four weeks old. The pathogen is also seedborne; however, the transmission of the pathogen to seedlings occurs most readily from freshly harvested or immature seeds, although systemically infected plants are usually sterile and produce no seed for harvesting. Seeds dried below 13% are unlikely to carry viable fungal material. Production of conidia requires the temperature of 17°C–29°C (optimum 24°C–26°C). Germination of conidia requires a saturated atmosphere or free water and moderate temperatures of 21°C–25°C. High levels of systemic infection occur at 11°C–32°C with a wet period of 4 hours or longer.

2.2.3 Philippine Downy Mildew

Symptoms

The characteristic greyish downy growth on leaves the followed by chlorotic discolourations, browning and necrosis of the blade and stunting of the plant. As the plant ages, leaves may narrow, become abnormally erect, and appear somewhat dried out. As the corn plant matures, tassels become malformed and produce less pollen, ear formation is interrupted, and sterility of seeds can result. If infection occurs early, plants are stunted and may die. There are no external symptoms on seeds.

Causal Organism and Etiology

Philippine downy mildew of maize is caused by *Peronosclerospora philippinensis*. The mycelia are branched, slender, irregularly constricted, and inflated. Erect conidiophores grow out of stomata and are dichotomously branched two to four times. Branches are robust. Sterigmata are ovoid to subulate, and slightly curved. The conidia are elongated ovoid to round cylindrical, hyaline, and slightly rounded at the apex. Haustoria are simple, vesiculiform to subdigitate.

Oospores are rarely produced and are not produced in corn tissue. When produced, oospores are spherical, and smooth-walled; they germinate by a side germ tube.

Disease Cycle and Epidemiology

Peronosclerospora philippinensis produce an over-wintering spore form (oospore), but its role in the life cycle has not been established. The pathogen has many alternate hosts such as "kans grass" *(Saccharum spontaneum)*, *Sorghum bicolor*, and *Sorghum halepens*. In the host, the mycelium produces the conidiophores, which bear the conidia. The conidiophores emerge through host stomata in the chlorotic areas on both leaf surfaces (more so on the lower surface of corn, because of the higher density of stomata), and sheaths, tassel rachis, glumes, and husks. The pathogen spreads intercellularly through the mesophyll cells. The fungus grows mainly downward through the leaf sheath to the stem where it moves into and persists in the shoot apex. When mycelium invades the meristematic tissues, chlorotic streaks soon appear on the leaves, followed by the fungus sporulating in these areas when conditions are favourable, producing secondary inoculum. The downy growth is the site of sporulation (conidia on conidiophores) and the source of the secondary spread of the disease to other susceptible plants. Germinating conidia produce germ tubes, which invade stomata. A mycelium develops in the mesophyll. Conidia form sparsely the morning after the first symptom appears, and abundantly after systemic symptoms appear. The disease spreads locally via wind and rain from an infected crop. Perennial grass hosts may also serve as reservoir hosts to carry over the pathogen during unfavourable periods or provide primary inoculum. The known infective agents are the mycelium in these plants or the airborne conidia produced by the mycelium. Although conidia are produced from 18°C–23°C and germinate from 10°C–35°C, the highest rates of infection occur at temperatures greater than 16°C. The conidial production, germination, and infection required night time temperatures of 21°C–26°C.

2.2.4 Sugarcane Downy Mildew

Symptoms

The most characteristic symptoms on maize are the development of long, rather broad, chlorotic strips along almost the entire length of the leaf. In some instances, apparently due to secondary infection, the stripes are short and narrow. Leaf shredding is not common. The downy growth of the pathogen can be seen on both surfaces of the leaf. Conidiation is more frequent on the lower leaf surface, but it can occur on both sides. Young leaves show the white hazy growth more prominently on older leaves. The pathogen also invades the shoot apex and the stem, but without producing any visible external symptoms. As the plant ages, the older leaves may become narrow, abnormally erect, and appear dry. Systemic infections can cause tassel malformation, interrupted ear formation, and reduction in pollen production.

Causal Organism and Etiology

Sugarcane downy mildew of maize is caused by *Peronosclerospora sacchari*. In the year 2013, the recently accepted taxonomic position of the genus *Peronosclerospora* is the following: kingdom–Chromista, phylum–Oomycota, order–Peronosporales, family–Peronosporaceae. Mycelia are wide (8 µm), irregularly constricted and inflated, growing intercellularly, in all the parts of the host plant except the root. Haustoria are simple, vesiculiform to subdigitate. Conidiophores are dichotomously branched 2 to 4 times; they develop through the stomata during the night dew or moist conditions. Conidia are borne on the sterigmata, hyaline, elongate ovoid to round cylindrical, slightly rounded at the apex. Oospores are regularly spherical and central to eccentric but form rarely.

Disease Cycle and Epidemiology

Peronosclerospora sacchari (syn. *Sclerospora sacchari*) survives as mycelium in sugarcane. Sporangia produced from mycelia at nightly temperatures of 20°C–25°C and in the presence of free water are disseminated by the wind to maize. Sporangia germinate in the presence of free water and form mycelia that penetrate through the stomata of plants that are less than one month old. Mycelia then develop systemically in the infected plant. Sporangia are eventually produced in diseased tissue and serve as secondary inoculum. Oospores are also produced, but their function is not known. Older plants are resistant. Other hosts are broomcorn, gamagrass, grain sorghum, sugarcane, and teosinte. The most intensive production and germination of conidia are influenced by temperature, humidity, and light. Both production and germination of conidia require a temperature of 23°C–25°C for 6–8 hours and relative humidity of 90%–94%. Free water on the leaf surface is essential formation and germination of conidia. In nature, the production of conidia mostly occurs during the night.

2.2.5 Rajasthan Downy Mildew

Symptoms

The typical symptom is *half diseased leaf*, characterised by the pale appearance of the bases of the leaves of the seedling. On infected leaves, yellow stripes at the base also extend up to the upper green portion. Severely infected plants give a yellowish appearance even from a distance. Most of the infected plants die at the about knee-high stage. Under humid conditions, whitish fluffy growth due to abundant fructification of the fungus can be observed on the lower and upper leaf surfaces. Systemically infected maize plants generally do not form cob.

Causal Organism and Etiology

Rajasthan downy mildew of maize is caused by *Peronosclerospora heteropogoni*. Conidiophores: Determinate, macronemous with swollen base, dichotomously branched. Conidia: Globose, hyaline, thin-walled. Oospore: Spherical with a persistent oogonial wall.

Disease Cycle and Epidemiology

P. heteropogoni produces abundant conidia and oospores on leaves of *H. contortus* which acts as a collateral host but produces only conidia on maize leaves. Oospores remain in the soil along with infected leaf and root residues and cause primary infection in the following years. Oospores formed on this grass are thus the primary source of inoculum which first infects the grass and then maize at the 2–3 leaves stage. No oospore formation by *P. heteropogoni* has been observed on maize. Conidia are produced on both *H. contortus* and maize and germinate instantly through the germ tube. Systemically infected leaves of maize produce abundant conidia on both surfaces. The conidia are the main source of the secondary spread of the disease within and among fields. The highest infections of *P. heteropogoni* occur in the temperature range of 23.5 ± 0.1°C with rainfall of 154.4 mm and relative humidity of 87.3%. Conidial germination occurred between 10°C–35°C, maximum at 25°C, and no germination occurred at 40°C.

2.2.6 Crazy Top

Symptoms

Generally, excessive tillering, rolling, and twisting of the upper leaves appear first before the malformation of tassels. The most conspicuous symptom is the proliferation of the tassel, instead of a normal tassel, a mass of narrow, twisted and distorted leafy structures develops. These modified leaflike inflorescences (phyllody) are called crazy tops. The affected plants show excessive tillering, severe stunting, narrow (strap-like) leathery leaves and long yellow semi-transparent, strips appear on "normal" leaves. This is a result of a reduced number of chloroplasts, later on, necroses follow. The tassels of infected plants do not produce pollen. Ear may also have excessive, elongated, leaf-like structures and are sterile.

Causal Organism and Etiology

Crazy top of maize is caused by *Sclerophthora macrospora*. It is a species of heterokont in the family downy mildews. Oogonia are globose to subglobose, thick wall, hyaline, 20–48 × 10–29 µm, producing oospore filling oogonia. Antheridia hyaline, kidney-shaped, paragynous. Sexual spores (resting spores or oospores) of the *Sclerophthtora macrospora* are hyaline to yellowish, lemon-shaped, multinuclei, relatively large, and have a thick wall. Oospores produce zoosporangia after germination. Zoosporangia ellipsoid to obovoid, hyaline, apical papillate, 60–114 × 28–50 µm, germinating in water to release kidney-shaped zoospores with 2 flagella. The size of the oospore of *Sclerophthora macrospora* is 60 to 65 µm, which makes them the largest among all the species of downy mildew on maize.

Disease Cycle and Epidemiology

The pathogen survives in infested corn residue, infected kernels, and wild grasses in the form of latent mycelium and oospores. Over 140 grass species are hosts of *S. macrospora*. The pathogen is also transmitted through seeds. The oospore germinates only after a period of dormancy of not less than 8 weeks. The zoospores attract to roots chemotactically and penetrate the plant behind the root cap. After germination, oospores form a thin-walled tube that bears a lemon-shaped sporangium. The sporangia germinate at 12°C–16°C, producing numerous motile zoospores. Zoospores are liberated in water-saturated soil and "swim" to seedlings. The zoospores encyst and produce a germ tube that penetrates seedling host tissue and becomes systemic. Sporangia are rarely produced directly on sporangiophores projecting from stomata on leaves, but numerous oospores are produced within diseased leaves and leaf sheaths. Intense rain, ponding, and prolonged flooding increase the risk for crazy top disease. Infection occurs over a wide range of soil temperatures.

Management

Cultural Methods

- Remove and destroy crop residue.
- Deep ploughing.
- Follow long crop rotation with pulses.
- Rogue out infected plants which can be used as fodder.
- Eradicate collateral and wild hosts.
- Provide adequate soil drainage.
- Do not grow maize in low and wet spots.
- To avoid, brown stripe downy mildew of maize, planting of maize should be done before the beginning of the rainy season.
- In case of sorghum downy mildew, avoid maize-sorghum crop rotation in the field and sowing of maize adjacent to a field of sorghum to avoid the spread of secondary infection
- Rogue and destroy infected plants and alternate host (*Hetropogon* grass) to manage the Rajsthan downy mildew.

Host Resistance

Grow resistant or tolerant maize varieties such as:

- For brown stripe downy mildew: PAU 352, Pratap Makka 3, Gujarat Makka 4, Shalimar KG 1, Shalimar KG 2, PEMH 5, BIO 9636, NECH-X 1280.
- For sorghum downy mildew: DMH 1, NAC 6002, COH (M) 4, COH (M) 5, Nityashree.
- For Rajsthan downy mildew: PEMH 5, Bio 9636, NECH-X 1280, etc.

Chemical Control

- Treat the seed with fungicide Metalaxyl 35% WS @ 0.75 kg to 1.0 kg/100 kg of seed in the endemic areas.
- Treat the seed with Ridomil MZ 75 WP @ 4 g/kg of seed.
- Foliar spray of Mancozeb 75% WP @ 1.5 to 2 kg/L at the very first appearance of symptoms of downy mildews.
- Spray Dithane M-45 (0.3%) + Zinc sulphate (0.05%) mixture.

2.3 LEAF SPOTS

2.3.1 Brown Spot

Brown spot of maize is common in several maize growing regions of the world, including India. It was first noticed in Bihar (India) by Shaw in 1910. It can cause 5%–30% yield losses.

Symptoms

Brown spot lesions first appear as very small, round-to-oblong, yellowish spots on the leaf blade, leaf sheath, stalk, and rarely on the husks and tassel of the outer ear. The spots may occur in bands across the leaf blade. Infected tissues turn a chocolate brown to reddish brown and merge to form large blotches with an irregular, angular appearance. Cells of infected corn tissue disintegrate to expose dusty pustules (brown blisters) containing enormous numbers of microscopic sporangia. The sporangia are a golden brown to dark brown. Nodes and internodes also show brown lesions. In severe cases of infection at the nodes beneath the leaf sheaths and the premature death of plants from leaf blights favour stalk rot and lodging.

Causal Organism and Etiology

Brown spot of maize is caused by an obligate parasite, *Physoderma maydis* (Teleomorph: *Cladochytrium maydis*). This is chytridiomycete fungus that produces coenocytic hyphae and yellow to brown thick-walled, flat, smooth brown sporangia (resting spores) produced terminally or intercalarily on rhizomycelium within infected cells. Endosporangium is conical or pyriform, thin-walled with saucer-shaped lid with apical papilla. Zoospore is posteriorly uniflagellate and ellipsoidal, with a large eccentric hyaline refractive globule.

Disease Cycle and Epidemiology

The fungus (*Physoderma maydis*) is an obligate pathogen, surviving inside the host tissue as sporangia even after harvest. The resistant sporangia are released from infected plants, disintegrating corn debris, and soil and are carried to susceptible plants by air currents, insects, splashing rain or irrigation water, and humans. Free water is required for infection. During the next cropping season, when sufficient moisture is present, the sporangia spread as airborne spores, releasing uniciliate (single flagella), hyaline, and thin-walled zoospores. The zoospores attached to the young leaves and germinate to produce infection hyphae, which later enter the host tissue and produce characteristic spots. Infection hyphae expand within the host cells to form several enlarged storage cells called *Sammelzellen*. High temperature (28°C–30°C) and abundant moisture during the early growth stage of the host favour disease development. When moisture is present in the whorl or behind the leaf sheaths and temperatures are relatively high (23°C–30°C), a sporangium "germinates" to release 20 to 50 swimming zoospores. The zoospores move about in the water for 1 to 2 hours before settling down, becoming amoeba-like, and penetrating young meristematic tissue with fine infection hyphae. The resulting mycelium enters mesophyll or parenchyma cells and forms larger vegetative structures. The development of symptoms and the germination of new sporangia occur approximately 6 to 20 days after infection, completing the disease cycle.

Management

Cultural Methods

- Field sanitation reduces the inoculum potential.
- Avoid planting highly susceptible hybrids.
- Avoid planting known susceptible varieties in highly humid areas.

- Crop rotation and tillage practices may reduce pathogen inoculum. As *Physoderma* is known to remain viable in the soil, as sporangia, for about 3 years.
- Eliminate crop debris from previous crops.

Host Resistance

- Grow resistant varieties such as JH 10655 and FH 3113.

Chemical Control

- Spray any copper fungicide (e.g. Copper oxychloride, etc.) @ 3 g/L of water.

2.3.2 Grey Leaf Spot

Since it was recognised as a *disease on the move* by Latterell and Rossi (1983), grey leaf spot has become increasingly important and is currently seen as one of the most serious yield reducing and widespread diseases of maize in the world.

Symptoms

Symptoms are typically first observed on the lower leaves. At the early stage appear on leaves as small (1–3 mm long), pinpoint lesions surrounded by yellow halos, but as lesions mature, they elongate into narrow, rectangular, brown to grey spots. Mature grey leaf spot lesions are grey to tan in colour and are distinctly rectangular (5–70 mm long and 2–4 mm wide). A distinguishing feature of grey leaf spots is that lesions typically run parallel with leaf veins. Sporulating lesions assume a greyish cast, hence the name grey leaf spot. Further lesion expansion results in the coalescing of lesions and the blighting of entire leaves. With severe blighting, stalk deterioration and severe lodging may occur.

Casual Organism and Etiology

Grey leaf spot of maize is caused by *Cercospora zeae-maydis*. The fungal genus *Cercospora* is one of the most ubiquitous groups of plant pathogenic fungi. Clustered conidiophores of *Cercospora zeae-maydis* arise from stomata on both leaf surfaces. Conidia are solitary, slightly curved, hyaline, 4–8 × 25–88 μm long with 3 to 5 septa.

Disease Cycle and Epidemiology

The disease cycle of *C. zeae-maydis* begins with conidia produced from the surviving pathogen in lesions on crop debris remaining on the soil surface. Pathgen is known to infect only maize and the pathogen is not seedborne. When conditions become favourable, the fungus produces conidia within infected crop residue. The conidia are wind-disseminated to newly planted maize crops. Lower leaves are usually the sites of primary infection in the developing canopy. *Cercospora zeae-maydis* infects maize leaves through stomatal pores, and light is required for the fungus to perceive stomata. Stomatal penetration, vegetative growth, and production of new conidia from stroma within the leaf lesion require 14 to 28 days. On nutrient-deficient substrates, but not on the water on the leaf surface, conidia germinate and develop secondary conidia on conidiophores produced from germ tubes or conidial cells. The new conidia are dessimenated by wind and rain to other leaves and plants, causing secondary infections throughout the growing season. *C. zeae-maydis* and many other *Cercospora* species produce cercosporin, a phytotoxin as a virulence factor. In the light, cercosporin generates active oxygen species that damage the plant cell membrane. Under favourable environmental conditions, lesions resulting from initial infections produce spores that are wind- or rain splash–disseminated to the upper leaves. It requires frequent and prolonged periods of high humidity and warm temperatures (20°C–30°C) to complete spore germination and the infection process.

Management

Cultural Methods

- Remove infected crop residue from previous crop.
- Use clean seed and resistant varieties.
- Follow crop rotation and destroy infected stubbles.
- Adopt tillage practices that aim to reduce initial inoculum by burying infected crop debris are classical methods and effective in managing the grey leaf spots of maize.
- Other practices such as time of planting, plant density, and timing of irrigation applications may all play a role in reducing disease severity.

Chemical Control

- Spry Copper oxychloride @ 3 g/L or Mancozeb @ 3 g/L or Carbendazim + Mancozeb 2 g/L of water.

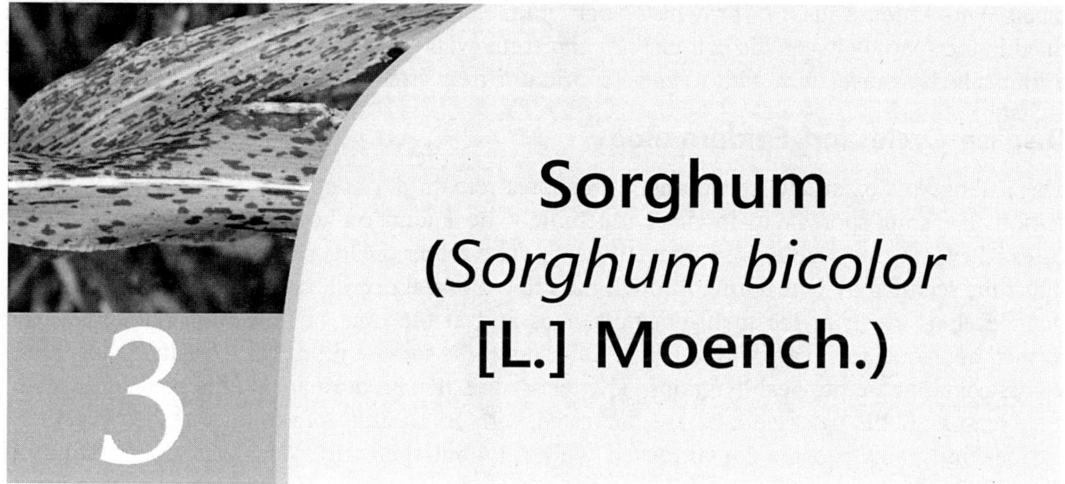

3.1 SMUTS

There are 7 smut diseases of sorghum, 4 of which occur in India:

- Grain smut
- Loose smut
- Long smut
- Head smut

3.1.1 Grain Smut

Grain smut is also known as covered or kernel or short smut. This is the most destructive smut, causing extensive damage to grain yield all over the country. Ratoon crops exhibit a higher incidence of disease.

Symptoms

The disease becomes apparent only at the time of grain formation. Most of the grains of an infected ear are replaced by the smut sori but in some cases only a few. The sori are oval to cylindrical, dirty-grey, and covered with a tough creamy skin (peridium) which often persists unbroken up to thrashing.

Causal Organism and Etiology

Sporisorium sorghi (syn. *Sphacelotheca sorghi*) causes covered kernel smut. The pathogen is present in the form of the sorus, which has a tough wall and a long, hard, central tissue, known as *columellum*. A dense mass of brownish-black, minutely echinulate thick-walled teliospores, are filled in the space between the columellum and sorus wall. These spores germinate and produce a four-celled promycelium with a single sporidium from each cell.

Disease Cycle and Epidemiology

The pathogen is externally seedborne. The spores remain dormant on the seed until the next season. The smut spores remain viable indefinitely, depending on how they are stored. In some cases, they may be viable even after 10 years. The teliospores germinate with the seed and infect the seedling by penetrating through the radical or mesophyll, establish systemic infection that develops along in the meristematic tissues and at the time of flowering, the spores are formed, replacing the ovary with the sori. Later on, these sori are ruptured, releasing teliospores, which contaminate the healthy grains. The teliospores remain dormant on the seed until sown next season. If the diseased ears are harvested with the healthy ones and threshed together, the healthy grains become contaminated with the smut spores released from the bursting of the sori. The incidence of smut decreases when seeds are planted during high temperatures.

Management

- Collect smutted ear-heads in cloth bags and destroy them by dipping in boiling water or bury in the soil.
- Avoid ratoon cropping.
- Grow resistant varieties.
- Treat the seed with Sulphur or Thiram @ 4.0 g/kg of seed.
- Treat the seed Sulphur 80% WP @ 3–4 g in one liter of water per 10 kg seeds or by Thiram 75% WS @ 25–30 g in one liter of water/kg seeds.

3.1.2 Loose Smut

The disease has been reported in most sorghum growing regions of the world. However, it is less common than covered kernel smut.

Symptoms

The affected plants are shorter, produce thinner stalks, more tillers, and earlier flowering than the healthy plants. Generally, all the spikelets in an infected panicle are smutted. Individual infected kernels are replaced by smut sori. They also develop on glumes and pedicel. The sori (3–18 × 2–4 mm size) are covered by a thin grey membrane that ruptures, exposing a powdery mass of dark coloured spores, before the emergence of panicle from the boot.

Causal Organism and Etiology

Sporisorium cruentum (syn. *Sphacelotheca cruenta*) causes loose smut of sorghum. As in covered kernel smut, at the center of the sorus, a columellum is present which is usually bigger and more curved than the columellum of covered smut. The spores are globose to sub-globose, dark brown, minutely echinulate and measure 5–10 μm in diameter. They germinate by forming a four-celled promycelium and sporidia. The sporidia multiply by budding.

Disease Cycle and Epidemiology

The pathogen is mainly externally seedborne. When the seed contaminated with teliospores is sown in the field, the spores germinate by producing hyphae, which infect young seedlings before emergence. Thereafter, the pathogen grows systemically and ultimately sporulates in the floral organs, resulting in smutted panicles. Seedling infection occurs at a varying soil moisture and temperature of 20°C–25°C. Since the disease is systemic, ratoon crops are also infected. Spores from a smutted panicle may infect late developing panicles of healthy plants thus causing a secondary infection. This infection is localised, and no subsequent systemic spread of the pathogen occurs.

Management

- Rogue out affected plants.
- Ratooning of the susceptible crop should be avoided.
- Treat the seed Sulphur 80% WP @ 3–4 g in one liter of water per 10 kg seeds or by Thiram 75% WS @ 25–30 g in one liter of water/kg seeds.

3.1.3 Long Smut

Long smut was first reported from Egypt in 1887, is now prevalent in many countries in Africa and Asia.

Symptoms

Long smut appears as elongated, cylindrical, slightly curved sori, much longer than normal grains. The sori are covered by a whitish thin membrane that ruptures to release black powdery mass of spores. Usually, a few smut sori are scattered sporadically throughout the panicle.

Causal Organism and Etiology

Long smut is caused by *Sporisorium ehrenbergii* (syn. *Tolyposporium ehrenbergii*). The pathogen produces characteristic sori filled with black masses of teliospores and also contains 8–10 dark filamentous structures of vascular tissues of the ovary arising from the basal portion of the sorus. Teliospores are firmly united into balls, light brown in colour. They germinate in water droplets and produce a 3–6 celled promycelium. The promycelium bears numerous sporidia at the septa and terminal ends. Sporidia are hyaline, spindle-shaped, and single-celled.

Disease Cycle and Epidemiology

The fungus survives as teliospores on the seed surface and in soil. The pathogen is also airborne and infects single florets. Under favourable conditions, the teliospores germinate and release numerous sporidia that may land in the boot leaf of plants and cause infection. Airborne spores may also produce sporidia on the flag leaf sheath which can infect the florets in the panicle. A temperature of 30°C–35°C and more than 80% RH favour disease development.

Management

- Adjusting the date of sowing.
- Destroy the infected heads.
- Eliminate possible alternative hosts.

3.1.4 Head Smut

Head smut of sorghum has been reported from many countries of the world. The disease causes significant damage as it affects the entire ear transforming it into a smutted head.

Symptoms

The disease appears at the time of flowering. In the place of a normal inflorescence, a sorus fully covered with a greyish-white membrane emerges from the boot leaf. The sorus is usually 3–4 inch long and 1–2 inches wide and maybe cylindrical in shape. When it has fully emerged the fungal membrane ruptures, exposing a large mass of black, powdery spores. If the wind is blowing during sorus emergence, the spores resemble a smoky cloud around the head. When the spores are blown off, a network structure of filamentous vascular tissues of the host is exposed. Diseased plants remain stunted, tiller excessively and the root system of the smutted plants is weakened.

Causal Organism and Etiology

Head smut of sorghum is caused by *Sporisorium reilianum* (syn. *Sphacelotheca reiliana*). Teliospores or smut spores develop from condensation of sporogenous hyphae into coils of dense cytoplasm, separated by partitioning hyphae. As the sorus matures, the dark, spherical, reticulate spores are liberated. Teliospores germinate and produce masses of sporidia.

Disease Cycle and Epidemiology

The young heads enclosed in the boot are completely replaced by large smut galls covered by a thick whitish membrane. The membrane ruptures, often before the head emergence, exposing a mass of dark brown to black, powdery teliospores intermingled with a network of long, thin, dark, filaments of vascular tissue. The pathogen is soilborne in the form of teliospores which germinate and cause infection in the nodal region of the shoot apex. Mycelium grows both intracellularly and intercellularly. During flowering, the mycelium grows vigorously and produces teliospores. Spores produced from infected plants reach the soil and survive until the next cropping season. Spores may also adhere to the seed surface and spread through the seed. Dry soil with a temperature of approximately 24°C is favourable for infection. Spores are known to survive for long periods in soil.

Management

- Collect all smutted ears in cloth bags and dip them in boiling water to reduce the inoculum potential for the next crop.
- Rouge out all smutted ears and burn.

- Follow crop rotation.
- Treat the seed with Captan or Thiram @ 4 g/kg seed.

3.2 GRAIN MOLD

Grain mold is one of the major constraints in sorghum production. Grain mold results in reduced seed size and weight, discolouration of grains, decreased germination, decreased seedling vigor, and mycotoxin contamination. The loss resulting from this disease ranges from 30% to 100% depending on cultivar type, flowering time, and prevailing weather conditions.

Symptoms

The initial symptom of the disease is discolouration of grains due to infection and colonization by mold fungi. Partially infected grains show whitish, greyish, orange, and pinkish to shiny black discolouration depending on infection by a particular fungal species. Fungal growth appears at the hilar end of the grain and thereafter extends to the pericarp surface. Often grains are colonized by many fungi. Heavily infected grains turn completely black.

Causal Organism and Etiology

Fungi associated with grain mold complex include *Fusarium* spp., *Curvularia lunata, Alternaria alternata,* and *Phoma sorghina*. Among Fusarium species *F. andiyazi*, *F. proliferatum* and *F. thapsinum* are proven pathogens of sorghum grain mold.

Fungi	Diagnostic features
Fusarium spp.	Generally produces pinkish white mycelium, powdery in appearance at first which later becomes pinkish fluffy.
Curvularia lunata	Shiny black, fluffy growth on the grain surface.
Alternaria alternata	Dull with greyish-black mycelium, often sparse and in stripes.
Phoma sorghina	Produces pin-like small, round, black pycnidia embedded in grain and produces a thick dirty black crust with a rough surface on the pericarp.

Disease Cycle and Epidemiology

Most of mold-causing fungi are weak parasites disseminated by seedborne, soilborne, and airborne spores. However, the role of seedborne inoculum as a direct cause of grain mold is minimal. Infection and colonization of flowers occur prior to grain maturity. Mycelium penetrates the pericarp and ramifies within 5–10 days. Fungus subsequently invades the endosperm and sometimes the embryo as well. At the time of anthesis, sorghum flower is most susceptible to infection and colonization by grain mold fungi. Early infection on the apical portions of flower tissues occurs on glumes, lemma, and palea and subsequently, grains are covered by fungal growth and sporulation. Spores are readily disseminated by wind and rain splash and spread the disease. Humid and warm conditions during flowering and grain development favour infection while dry conditions prevent it. Spore production increased in warm temperature (25°C–28°C) and high relative humidity (100%) and decreased with the drop in temperature below 15°C and rise in temperature above 30°C. With a sudden increase in the relative humidity following rainfall, the inoculum load increases several times.

Management

- Adjust planting dates so that plants do not face frequent rains during the grain filling stage.

Host Resistance

- Grow resistant or tolerant varieties. Many grain mold tolerant lines are GMRP 4, GMRP9, GMRP13, GMRP25, GMRP28, GMRP33, ICSV91008, ICSV95001) and superior mold tolerant cultivars that are widely cultivated in India are CSH16, SVD9601, PVK801, CSH9, and SPV462.

Biological Control

- Treat the seed with *Pseudomonas fluorescens* formulation to reduce *F. moniliforme* infection in sorghum.
- Give two sprays of talc-based formulation of Fluorescent Pseudomonad @ 0.2% at flowering and 20 days after flowering.

Chemical Control

- Treat the seeds with Chloranil or Thiram (3 g/kg seed) and Bavistin (2.5 g/kg seed).
- Give 3 foliar sprays of Mancozeb + Captan @ 0.2% at flowering, 15, and 25 days after flowering or give a foliar spray of Captafol @ 0.2% or Carbendazim + Triadimefon or spray of Propiconazole 25 EC @ 0.1% at flowering or give 3 sprays of Thiram + Carbendazim @ 0.2% or Captan + Mancozeb @ 0.3% at flowering, 15, and 25 days after flowering.

3.3 ANTHRACNOSE

Anthracnose is one of the most destructive diseases of sorghum throughout the world. Leaf blight phase is the most devastating and may result in yield losses of 50% or more under severe conditions. The pathogen causes seedling blight, leaf blight, stalk rot, head blight, and grain molding thus limiting both grain and forage production.

Symptoms

Foliar symptoms usually appear 25–30 days after emergence. Symptoms are characterised by small, circular to elliptical spots, up to 5 mm in diameter but often smaller, which develop grey to straw-coloured centers with wide purple or red margins depending on the host cultivar (Figure 3.1). Under hot and humid conditions, the spots increase in number and coalesce to cover large leaf areas. In the center of the spots, small, circular, black dot-like fruiting bodies (acervuli) develop. These acervuli, with setae (small, black, hairlike structures protruding from the acervuli), are diagnostic characteristics of anthracnose. Midrib infection often occurs and is seen as elliptical to elongated red or purple lesions on which the black acervuli can be clearly seen. The use of a hand lens or a dissecting microscope helps identify these structures. Leaf-sheath and panicles including grains and rachis are also infected. Symptoms of leaf anthracnose usually become clearly visible about the time of boot formation. These symptoms are sometimes easier to recognise on the mid-vein of older leaves. The disease can also infect the stalk of grain sorghum hybrids.

Figure 3.1 Anthracnose of Sorghum.

Causal Organism and Etiology

Anthracnose of sorghum is caused by *Colletotrichum graminicola*. Conidia are produced terminally on the erect, nonseptate, and short conidiophores among the setae. They are hyaline, nonseptate, uninucleate, and cylindric to obclavate but become sickle-shaped. Acervuli produced on the infected host tissue are dark brown, oval to cylindrical and with or without setae.

Disease Cycle and Epidemiology

The pathogen (*Colletotrichum graminicola*) survives on host residue, wild sorghum species, and weeds as conidia or mycelium on the seed. It can persist for more than a year in diseased residues on the soil surface. Conidia from wild sorghum species or residues serve as the primary inoculum carried to sorghum leaves by wind or splashing rain. Conidia germinate and penetrate the epidermis directly or through stomata. Seedling infection occurs from the inoculum present in the crop residue. The disease is particularly severe in warm and humid environments resulting in substantial economic losses. The disease becomes most severe during the period of continuous rain, high humidity, and temperature of 28°C–30°C.

Management

- Collect and destroy disease infected plant or its parts.
- Clean cultivation and elimination of grasses have been used to manage the disease.
- Eliminate other susceptible plants such as Johnson grass.
- In poorly rotated fields, plough under infested residue where erosion is not a problem.
- *Trichoderma* spp. and fluorescent *Pseudomonads* have shown potential to be used as a biocontrol agent for sorghum anthracnose.

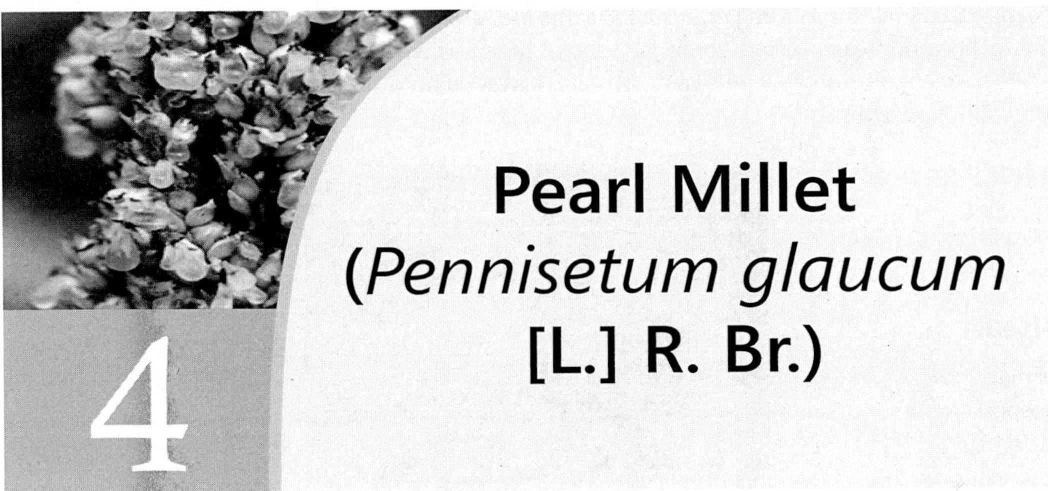

Pearl Millet (*Pennisetum glaucum* [L.] R. Br.)

4.1 DOWNY MILDEW

Downy mildew is the most common disease in all areas where bajra is grown. Loss estimates in different countries vary from 6%–60%. In India, it can cause losses of upto 27%–30% and upto 60% in African countries.

Symptoms

Typical downy mildew symptoms include chlorosis, sporulation on the lower leaf surface, stunted growth, and malformation of the earheads (green ear). However, the symptoms can be divided mainly into two groups: *downy mildew* and *green ear*.

Downey mildew symptoms: Symptoms on the leaf initially appear as chlorosis (yellowing) at the base of the leaf lamina, and successively younger leaves show a progression of chlorosis. White downy growth appears on the lower surface of an infected chlorotic leaf due to abundant white asexual sporulation (spornagiophores and sporangia). This surface growth is more profuse during rainy and humid conditions. Subsequently, the leaves turn reddish brown and dry, due to oospore production. The *half-leaf symptom* is a characteristic symptom of the disease which is shown by a distinct margin between the diseased (basal) portion and the non-diseased areas toward the tip of the leaf. Severely infected plants are generally stunted because of root retardation and do not produce panicles.

Green ear symptoms: This symptom appears on panicles due to the transformation of floral parts into leafy structures. Ear deformities are characterised by the transformation of all floral organs into twisted leafy structures. This gives ear the appearance of a green leafy mass hence named *green ear*. Such panicles generally do not produce grain. The leafy structures are chlorotic and sometimes produce sporulation.

Causal Organism and Etiology

Downy mildew of pearl millet of *Sclerospora graminicola*, an Oomycetes (fungi-like organism) causes green ear disease of bajra. It is a biotrophic pathogen and produces small, bulbous haustoria to obtain nutrients from infected host tissues. The short, stout, hyaline sporangiophores of this pathogen emerge in clusters through stomata. The sporangia are hyaline, thin-walled, and elliptical, and bear prominent papilla. They germinate in presence of water, releasing 3–12 zoospores. The oospores can be seen in shredded host tissues. Oospores are the sexual spores of *S. graminicola* and are brownish-yellow, thick-walled, spherical, and resting spores.

Disease Cycle and Epidemiology

Primary source of infection of *Sclerospora graminicola* are soil and systemically infected seeds. Oospores, present in the soil, serve as the primary source of inoculum and infect the underground parts of plants, mostly at the seedling stage. Wild species of *Panicum* also play a role in the perennation of the pathogen and its variability. *P. violaceum*, *P. millissium*, *P. purpureum*, *P. pedicellatum* and *P. polystachyon* are susceptible wild or collateral hosts. *S. graminicola* oospores mostly germinate directly by germ tubes. The infection process begins with the formation of a germ tube which produces an appressorium. Systemic symptoms appear when the pathogen invades the developing leaves or inflorescence at the growing point. Sporangia are important for the secondary spread of the disease within and among fields if environmental conditions are suitable. Dispersal of sporangia takes place through wind, water, and insects. These sporangia release zoospores which germinate, penetrate the epidermis or stomata and causes infection in successive manner. Disease development is favoured by high relative humidity (85%–90% RH) and moderate temperature (20°C–30°C). *S. graminicola* survives as oospores in the soil along with infected leaf residue, and cause primary infection in the subsequent years. Oospores are transmitted on the seed surface, in soil, by wind, or by water.

Management

Cultural Methods

- Use of disease-free seed.
- Burn downy mildew-infected plant debris. Effective reduction of oospore-infested debris after harvest of the crop is a prerequisite to reduce primary inoculum for the subsequent crops.
- Deep ploughing of the field should be done to bury the infected crop debris.
- Do early showing of pearl millet crop in the season. Generally, very early showing of pearl millet has a chance to escape the infection from *S. graminicola* than that sown late in the season due to less inoculum (sporangial) buildup and low infection by soil-borne oospores.
- Follow crop rotation with nonhost crops. Oospores of *S. graminicola* survive in soil and are considered to be the primary source of infection; therefore, crop rotation can be another disease management strategy. Crops such as wheat, sorghum, sugarcane, and black gram appeared to increase downy mildew incidence while cotton, coriander, and onion drastically reduced the disease incidence.

- Removal (rouging) and complete destruction of infected plants from the field. This can reduce the secondary spread of disease during the same season.

Biological Control

- Treat seed treatment with talc formulation of *Trichoderma harzianum* @ 20 g/kg of seeds and talc formulations of *Pseudomonas fluorescens* and *Bacillus* species @ 10 g/kg of seeds.
- Seed treatment with many biopolymers derived from plants like *Acacia arabica,* neem, drum stick, papaya, atrocarpus and mimosops are highly effective in managing pearl millet downy mildew at a concentration of 1:2 w/v in combination with half dosage of Metalaxyl @ 3 g/kg of seed.

Chemical Control

- Seed treatment with Metalaxyl 35% WS @ 6 g/kg of seed controls the disease excellently for about the first 35 days after sowing.
- Foliar application of the Ridomil MZ 72 @ 3 g/L of water arrests further development of the disease in systemically infected plants.
- Several new generation chemicals have been found effective in the control of downy mildew such as the safe and eco-friendly strobilurin group of fungicides/oomyceticides-Trifloxystrobin, Azoxystrobin and Kresoxim-methyl, and Cyazofamid and Iprovalicarb.

4.2 ERGOT

The ergot of pearl millet was first reported from South India but was not considered a major disease of the crop. This disease became a major yield constraint in India with the introduction of open-pollinated hybrid lines of pearl millet. Ergot infection can have a significant impact on yields, with up to 70% loss in susceptible varieties. Ergot disease is important because it adversely affects grain quality by contaminating the grain with a neurotoxic alkaloid contained in the sclerotia.

Symptoms

The first symptom of ergot appears as cream to pink mucilaginous droplets of *honeydew* oozing out from infected florets on pearl millet panicles. Honeydew contains asexual spores (conidia) of the ergot fungus. Within 10 to 15 days, the droplets dry and harden, and dark brown to black sclerotia develop in place of seeds on the panicle. Sclerotia, commonly referred to as an ergot, are larger than seeds and irregular in shape. They are often elongated and protrude from the glumes of maturing heads. During harvesting and threshing, these sclerotia fall on the ground and generally get mixed with the grains and serve as a source of primary inoculum for the next crop. These sclerotia contain alkaloids responsible for the poisoning in animals. If these alkaloids are ingested they can result in convulsions, hallucinations, gangrene, and death.

Causal Organism and Etiology

Ergot of bajra is caused by *Claviceps fusiformis*. The honeydew produced on the ear is full of conidia and consists of two types of asexual conidia (macro and micro-conidia). The macroconidia are hyaline, fusiform, and broadly falcate and unicellular. Microconidia are globular, hyaline, and unicellular. The honeydew stage is followed by the development of sclerotia which are elongated to round, light to dark brown, and hard to brittle, with cavities filled with conidia. Sclerotia germinate to form (1 to 16) fleshy, 6 to 26 mm long purplish stipes. Each stipe bears a globular capitulum or perithecial stroma at the tip. The stroma is light to dark brown, with numerous perithecial projections. The perithecia contain asci which are interspersed with paraphyses. The asci contain thread-like, hyaline, and aseptate ascospores (sexual spores).

Disease Cycle and Epidemiology

The primary disease cycle of *Claviceps fusiformis* begins with the sclerotia left in the field during harvest or mixed with the seed at the time of threshing and sown along with the seed the next season. Ergot sclerotia from the infected panicles fall to the ground at harvest, or get mixed with the seed during threshing and serve as a primary source of inoculum for the next crop. Sclerotia germinate following rain. The carpogenic germination of sclerotia results in the production of perithecia and ascospores. The ascospores are dispersed by winds to flowering pearl millet panicles and cause primary infection. These ascospores germinate and infect the florets through the emerged stigma before pollination-fertilization. Honeydew production promotes secondary infection caused by conidia (asexual spores). The honeydew droplets contain numerous conidia of the pathogen. Conidia spread from plant to plant by wind, rainsplash, insects and physical contact between the diseased and healthy flowering panicles. Houseflies are very efficient in the dissemination of conidia from honeydew. On the same ear, honeydew trickles down to healthy florates and initiates infection. Within 10–15 days, honeydew droplets containing mycelial mass, conidia, and sugary liquid dry out and transformed into hard, dark brown to black structures called sclerotia. These are generally larger than the seed and vary in shape and size. Infection and disease spread are favoured by an overcast sky and drizzling rain that provides high humidity (more than 80%), moderate temperature (20°C–30°C), and wind during the flowering stage of pearl millet.

Management

Cultural Methods

- Use of clean seed for sowing. It is the most common recommendation as the sclerotia can contaminate the seed lots. Separate the sclerotia and its fragments effectively from the seed by flotation technique using a 10% NaCl solution. Dip the seeds in 10% NaCl salt solution for 10 minutes. Sclerotia can be removed by hand from small individual seed lots. Destroy the separated sclerotia by burning and washing the seed in clean water and drying under shade.
- Follow long (2 year) crop rotation in ergot-infested fields to avoid soilborne inoculum.
- Repeated deep ploughing during summer may reduce the viability of sclerotia.
- Intercropping of pearl millet with mungbean should be done to reduce disease incidence.
- Do early sowing (latest by 15th July) of pearl millet as it escapes the high disease incidence.

- Rouge out the ergot infected ears with honeydew and destroy by the burning.
- Remove infected crop residue from the field and destroy.
- Eradicate the weeds like *Cenchrus ciliaris* and *Panicum antidotale* from around pearl millet fields.

Host Resistance

- Use resistant varieties. It is the best and mosy cost-effective method of managing ergot disease in pearl millet.

Chemical Control

- After removal of sclerotia from contaminated seed, treat these with a fungicide like Thiram @ 2 g/kg of seed.
- Spray fungicides such as Benlate @ 0.2% at heading or stigma exsertion stage or Triadimenol @ 0.125 kg a.i./h or Propiconzole, Tebuconazole, Triadimenol, Myclobutanil or Azoxystrobin (25 µg/L) on full bloom panicles stage or a combination of Acibenzolar-S-methyl (0.05 kg a.i./h) and Mancozeb (1.5 kg a.i./h).

5 Groundnut (*Arachis hypogaea* L.)

5.1 EARLY AND LATE LEAF SPOTS

Leaf spot disease is commonly also known as *tikka disease*, *peanut cercosporosis*, *Mycosphaerella leaf spot*, *brown leaf spot* and *viruela*. Leaf spots are the most serious disease of groundnut in India and world-wide. Severe infection leads to defoliation and causes losses of over 50% in many parts of the world.

Symptom

Two types of fungi that cause leaf spots on groundnuts: early leaf spot and late leaf spot. They look very similar and are often found together even on the same plant (Figure 5.1, Table 5.1). Both types of fungi produce dark spots, roughly circular and up to 10 mm in diameter on the leaves. The presence or intensity of a yellow halo on the upper leaf surface is not always a reliable characteristic for identifying early and late leaf spots. Both fungi produce oval spots on stems, leaf stalks, and pegs, and cause leaves to fall off. The most reliable method of distinguishing between the two leaf spot diseases is to closely examine spots for the development of reproductive structures (sporulation). When spots are actively sporulating, the grey-coloured tufts of mold are visible with a 10X hand lens on the upper leaf surface for the early leaf spots and on the lower leaf surface for the late leaf spots. When defoliation is severe, lesions appear on stems, petioles, and pegs as dark brown to black, oval-shaped blotches.

Causal Organism and Etiology

Early leaf spot is caused by *Cercospora arachidicola* (sexual stage: *Mycosphaerella arachidis*) and late leaf spot is caused by *Cercosporidium personatum* (sexual stage: *M. berkeleyi*). These fungi exist mostly in their asexual states. *C. arachidicola* grows initially with intercellular and then intracellular when the host cells start to die. No haustoria are produced. Olivaceous brown

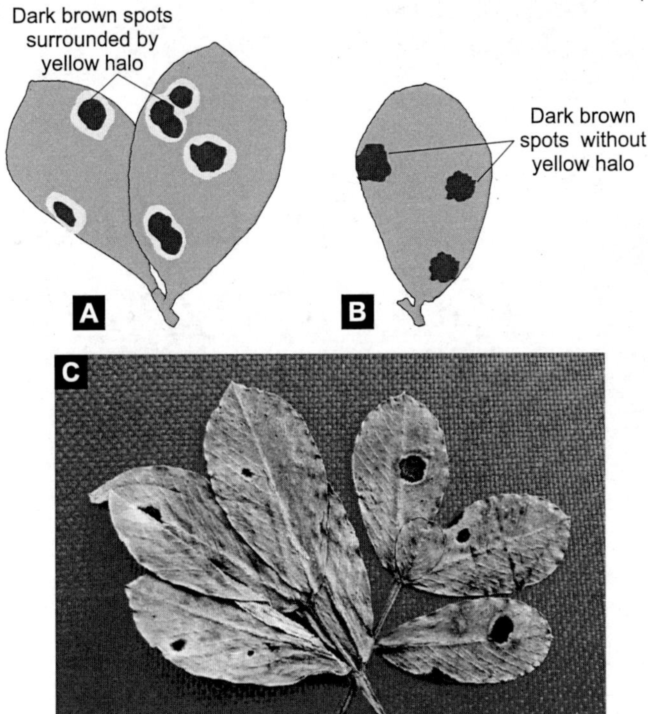

Figure 5.1 Tikka Disease of Groundnut. (A) and (C) Early leaf spot caused by *Cercospora arachidicola,* and (B) Late leaf spot caused by *C. personatum.*

Table 5.1 Difference between early and late leaf spots of groundnut

Features	Early leaf spot	Late leaf spot
Causal organism	*Cercospora arachidicola* (Sexual stage: *Mycosphaerella arachidis*	*Cercosporidium personatum* (Sexual stage: *Mycosphaerella berkeleyi*)
Symptoms	Spots are circular, larger, reddish-brown with a distinct yellow halo.	Spots are circular, small, dark brown to black; usually yellow halo is absent, if present it is smaller.
Location of spots on leaves	Upper and lower surface.	Mostly on lower surface.
Number of lesions	Less	More
Disease development	Rapid but causes less damage.	Slow, but causes more damage.
Sporulation of fungus	Mostly on upper surface.	Mostly on lower surface.
Concentric rings on spots	Not obvious	Visible on spots on the lower leaf surface due to the development of fruiting structures (spores) and their arrangement in concentric circles.
Seasonal development	Early	Late
Crop losses	Less	More

conidiophores are produced, continuous or 1–2 septate. Hyaline or pale yellow conidia, obclavate with rounded to distinctly truncate base and sub-acute tip. Septate mycelium is produced by *C. personatum* in the host and intercellular sending haustoria into the palisade and mesophyll cells. Conidiophores are uniformly olivaceous brown, continuous, sometimes 1–2 septa are present, without branches and geniculate. Conidia are cylindrical, light-coloured, 1–7 septate with bluntly rounded ends.

Disease Cycle and Epidemiology

The groundnut leaf spot fungi survive through conidia lying in the soil on diseased plant debris and conidia being carried on the shells of groundnut. The pathogens may also survive from season to season on volunteer groundnut plants and infected crop debris. Long-distance dispersal of the pathogens may be through airborne conidia, by the movement of infected crop debris, or by the movement of pods or seeds that are surface-contaminated with conidia. There are no reports on the internally seedborne nature of these pathogens. The disease onset is earliest and the attack is most severe where groundnut-groundnut cropping pattern is followed. Both diseases may appear within 3–5 weeks after sowing. The first lesions normally develop on the oldest leaves near the soil surface and the conidia produced on them are carried by wind, rain-splash, and insects to the later-formed leaves and to adjacent plants. The infection of *C. arachidicola* (early leaf spot) is followed by *Cercosporidium personatum* (late leaf spot). Although the perithecial stage of both species is known, but the ascospores are not generally regarded as important sources of primary inoculum. Conidia are produced directly from mycelium in crop debris in the soil following early rains and when deposited on the leaves of young groundnut plants by rain splash and wind, they initiate the disease cycle. Temperatures in the range 25°C–30°C, high relative humidity (more than 80%) and prolonged leaf wetness hours, favour infection and disease development.

Management

Cultural Methods

- Remove plant debris, weeds, and self-grown plants of groundnut.
- Follow crop rotation with the non-host crop.
- Bury crop debris deeply to destroy the soilborne inoculum.
- Mix crop with arhar.
- Do early sowing of groundnut.
- Use of early maturing varieties.
- Avoid sprinkler irrigation as it promotes disease spread.
- Pruning can also be done with the help of pruning scissors by removing the infected parts of the plant.

Biological Control

- Treat the seeds with talc-based formulation of *Pseudomonas fluorescens* @ 10 g/kg of seed.

Host Resistance

- Use resistant varieties. See Table 5.2 for important resistant/tolerant varieties of groundnut against leaf spots.

Table 5.2 List of some important resistant/tolerant varieties of groundnut against leaf spots

Disease	Resistant/tolerant varieties
Early leaf spot	Jawahar Groundnut-23, Kadiri-6, Kadiri Harithandhra, Pratap Mungphali-1, Pratap Mungphali-2, Pratap Raj Mungphali, Smruti, VRI-2, BAU-13, HNG-69, ICGV-00348, Kaushal, Amber, Chitra, Jawahar Groundnut-3, Kadiri-5, Vemana, Kadiri-8, M-522, M-548 and Somnath
Late leaf spot	Girmar-2, CG-7, GPBD-5, ICGV-00350, ICGV-91114, Jawahar Groundnut-23, Pratap Mungphali-1, Pratap Mungphali-2, Pratap Raj Mungphali, Smruti, TG-26, TG-37A, TMV (Gn)-13, Vikas, VRI-2, VL-Moongphali-1, BAU-13, ICGV-00348, Amber, Abhaya, Apoorva, ALR-3, Co (Gn)-4, Jawahar Groundnut-3, Kadiri Harithandhra, Kadiri-4, Kadiri-5, Kadiri-6, Kadiri-7, Kadiri-8, Phule Unap, Prutha, Ratneshwar, Vemana, VRI (Gn)-5, VRI (Gn)-6, VRI (Gn)-7, AK-265, Co 6, M-548 and Somnath

Chemical Control

- Disinfect the seeds of groundnut that are within the shells with Sulphuric acid and those which are without the shell can be treated with 0.5% of Copper sulphate solution for half an hour.
- Seed treatment and spray with suitable fungicides like Benlate are useful to avoid infection of the crop.
- Treat the seeds with Carbendazim or Thiram @ 2 g/kg.
- Spray Carbendazim @ 500 g/ha or Mancozeb @ 2 kg/ha or Chlorothalonil 2 kg/ha and if necessary, repeat after 15 days.
- Spray a mixture of Carbendazim 0.05% + Mancozeb 0.2% at 2–3 weeks intervals, 2 or 3 times starting from 4–5 weeks after planting.
- Spray Tebuconazole 25 EC @ 0.1% or Bitertanol 25 WP @ 0.1% at 15 days intervals.
- Azoxystrobin and Pyraclostrobin are more recent Strobilurin fungicides and have been found effective against leaf spots.

5.2 WILTS

5.2.1 Fusarium Wilt

Fusarium wilt is distributed worldwide. The fungus causes severe yield loss in groundnut. Groundnut wilt caused by *Fusarium oxysporum* was first reported in Tansania by Armstrong et al. (1975).

Symptoms

The disease normally occurs in plants subjected to prolonged drought stress. Infected plants may wilt suddenly or gradually. During sudden wilting, all leaves of entire plants turn greyish-green and the taproots show vascular browning. In dry weather, the entire plant becomes bleached and dry. When slow wilting occurs the foliage becomes chlorotic, and leaflets are shed before the plant dies. There are no external symptoms on either the stem or roots, but if the roots are cut longitudinally, vascular discolouration is evident. This discolouration is a common feature of other vascular wilts caused by *Verticillium* spp. and *Ralstonia solanacearum*. Hence, field diagnosis of fusarium wilt must be confirmed by laboratory tests. Pegs and pods may be invaded by the fungus which causes a pink discolouration of the inner pod surface.

Causal Organism and Etiology

Fusarium oxysporum and *Fusarium solani* causes wilt disease in groundnut. *F. oxysporum* is soilborne pathogen that causes Fusarium wilt disease in over hundred plant species. *Fusarium oxysporum* is a root pathogen colonizing the xylem and blocking them completely to effect wilting.

Disease Cycle and Epidemiology

Fusarium causes vascular diseases in plants such as watermelon, cucumber, tomato, pepper, muskmelon, bean, cotton, and groundnut. It invades roots and causes wilt diseases through colonization of xylem tissue. It causes a complete loss in plant grain yield if the disease occurs in the vegetative and reproductive stages of the crop. Due to the growth of the pathogen *Fusarium oxysporum* water supply in the plant's vascular tissue is affected which induces the stomata to close, the leaves wilt, and the plant eventually dies.

Management

- Seed treatment with systemic fungicides like Carbendazim @ 2 g/kg seed.

5.2.2 Verticillium Wilt

Symptoms

Early symptoms usually appear at the flowering stage. They include marginal chlorosis of leaves, loss of leaf turgidity, and leaf curling. Later symptoms are general yellowing and leaflet necrosis, followed by wilting and defoliation. During the early stages of disease development, plants wilt during the middle of sunny days but usually recover turgidity during the night. Wilting eventually becomes permanent. Wilt symptoms are generally more severe on younger plants than on older ones. The roots of infected plants have a brown discolouration of the vascular tissues. Occasionally plants die, and the roots of dead plants are severely rotted.

Causal Organism and Etiology

Verticillium wilt of groundnut is caused by *Verticillium albo-atrum* and *Verticillium dahlia*. Morphological features of *V. albo-atrum* and *V. dahlia* are given in Verticillium wilt of tomato.

Disease Cycle and Epidemiology

The fungus persists in soil for long periods as tiny seed-like structures called microsclerotia. Fields become contaminated with Verticillium where microsclerotia have been introduced with soil or water. Microsclerotia germinate and infect nearby roots. The fungus then grows into the vascular system resulting in plant wilting. Drought stress accelerates symptom development. Significant yield losses result when plants are killed prior to maturity.

Management

- Seed treatment with systemic fungicides like Carbendazim @ 2 g/kg seed.

5.2.3 Bacterial Wilt

Bacterial wilt caused by *Ralstonia solanacearum* species complex is an important soil-borne disease worldwide that affects more than 450 plant species, including peanut, leading to great yield and quality losses. Bacterial wilt, or slime disease, of peanut, was first observed in the East Indies in 1905. In China, the first report of an outbreak of bacterial wilt on peanut was in the 1930s. It is a major constraint to groundnut production in China, Indonesia, and Vietnam. Yield losses of 10%–30% commonly occur and can reach over 60% in heavily infested fields.

Symptoms

Infection of young plants can result in sudden wilting and death, but the leaves remain green. Infection of mature plants results in loss of turgidity, and leaves become light green chlorotic and curl at the tips. Eventually, leaflets become brown but remain attached to the plant. In some instances, only a single branch may wilt and die. The vascular system of the tap root becomes plugged and discoloured and this extends into the main stem and lateral branches. Bacterial ooze may be seen on the root, stem, and lower branches, and the oozing becomes evident as streaks of brown or black discolouration. The affected tissues then become black and show necrosis. Masses of bacterial ooze from the cut end of infected roots and stems when these are placed in water. When young plants are infected, the pods may remain small, or pods of such plants become wrinkled and may show rotting. When the mature plants are infected, there is no evidence of invasion of the pods.

Causal Organism and Etiology

Bacterial wilt of groundnut is caused by *Ralstonia solanacearum* (Burkolderiaceae, beta-proteobacteria). It is an aerobic, non-spore-forming, rod-shaped, gram-negative bacterium with a cell length of 0.5–1.5 µm. Virulent isolates are mainly non-flagellate and non-motile. Avirulent isolates usually bear 1–4 polar flagellae and are highly motile. Common fimbriae are often present in both virulent and avirulent isolates. *R. solanacearum* forms a highly diverse species complex encompassing four phylotypes, five races, and six biovars that have geographically distinct distribution.

Hosts Range: *Arachis hypogaea* (groundnut), *Ipomoea batatus* (sweet potato), *Gossypium hirsutum* (Cotton), *Capsicum* spp. (chillies), *Lycopersicon esculentum* (tomato), *Musa* spp. (banana), *Manihot esculenta* (cassava), *Solanum melongena* (brinjal), *Solanum tuberosum* (potato), *Nicotiana* spp. (tobacco), and *Zingiber officinale* (ginger).

Disease Cycle and Epidemiology

Ralstonia solanacearum overwinters in the soil. Infested soil and crop residues are the most important primary source of inoculum. The pathogen is mainly disseminated through infested soil, water, farm implements and machinery, cultural practices, even animals, and insects.

Dissemination through the seed is less important as the pathogen is highly sensitive to desiccation. The pathogen invades the groundnut plant through wounds or natural openings in roots. After the infected plants have wilted the bacteria are returned to the soil and can infect adjacent plants. The establishment, development, and disease severity of groundnut bacterial wilt are generally influenced by climatic factors, especially temperature and rainfall, soil type, pathogen population in soil, and host plant resistance. When the daily air temperature is over 20°C and the soil temperature in the top 5 cm layer is over 25°C for one week, symptoms of wilt in infested fields will occur. When the air temperature is over 25°C and the soil temperature over 30°C, the infection of the pathogen and wilt symptoms will reach their peak.

Management

Cultural Methods

- Follow crop rotation.
- Weeding and soil disinfection might extenuate bacterial wilt because contaminated weeds and soil are the major sources of *Ralstonia* spp.
- Soil amendment is widespread means used in soil management. For instance, the formulated product S-H mixture (4.4% bagasse, rice 8.4%, 4.25% oyster shell powder, urea, 8.25%, 1.04% potassium nitrate, 13.16 and 60.5% silicate slag SSP) is effective to control several soilborne diseases including fusarium wilt and bacterial wilt of many crops by enhancing the fertility and microbial abundance in soil.
- Calcium amendments like CaO and $CaCO_3$ are effective in managing bacterial wilt by inhibiting pathogen survival through changes in the pH and nitrite accumulation in the field.
- Organic fertilizer and biochar amendments are promising alternatives to suppress bacterial wilt by increasing the soil pH, electric conductivity, organic carbon and nitrogen availability, and microbial.

Host Resistance

Breeding of crop cultivars with suitable resistance is regarded as a key approach for the integrated management of bacterial wilt.

- Grow resistance varieties if available.

Biological Control

Biological control is another promising way to reduce bacterial wilt severity.

- The most frequently applied microbial agents are *Streptomyces* spp., *Bacillus* spp., *Pseudomonas* spp., avirulent *Ralstonia* spp. mutants, phage, and other microbes.

Chemical Control

- Soil fumigants like Chloropicrin, Dazomet, and Bromomethane can be applied to control bacterial wilt and other soil-borne diseases. However, these chemicals have been banned.
- Soil drenching with Copper oxychloride @ 3 g/L is also effective against bacterial wilt.

6 Soybean (*Glycine max* [L.] Merril.)

6.1 RHIZOCTONIA BLIGHT

Aerial blight, also called aerial web blight or Rhizoctonia foliar blight or Rhizoctonia web blight is a common disease on soybean in rice growing regions. In India, it was first reported from Pantnagar (Uttarakhand) in 1967. Extensive yield losses (40%–50%) have been reported in soybean when conditions favour disease development.

Symptoms

The pathogen may infect leaves, pods, and stems at all stages of development through maturity. Disease symptoms first appear on the lower portion of the plant growing in the closure canopy. Foliar symptoms often occur during late vegetative growth stages on the lower portion of the plant. Infected plants showing symptoms include leaf spots, leaf blighting, or defoliation. Initially, leaf symptoms appear as small, circular, water-soaked, light greenish-brown spots appearing on the leaf lamina. The spots increase in size, finally becoming oblong or irregular that turn tan to brown at maturity with slight reddening of the veins on the lower surface. In older spots, a dark pinkish-brown margin around the light tan-brown central necrotic tissue is common. Diseased tissue in old lesions that are no longer expanding generally falls out during dry weather, creating a ragged and shot-hole effect. Later the spots may coalesce and cover major or entire portions of the leaf. Such leaves turn yellow, dry, and finally drop off prematurely. Infected leaves droop and may adhere to pods and stem, thus becoming a source of infection for pods and seeds. Reddish-brown lesions can form on infected petioles, stems, pods, and petiole scars. Symptoms on petioles and stems are light to dark-brown, oblong, 1–2 mm wide, and up to 3–4 cm long. Lesions on pods may be small, brownish spots or may blight the whole pod. Seed infection is associated with pod infection. Pods and stem tissues that are infected will be greasy brown and shriveled. Under field conditions, a web-like mycelium with micro- and/or macrosclerotia may form on plant surfaces (Figure 6.1).

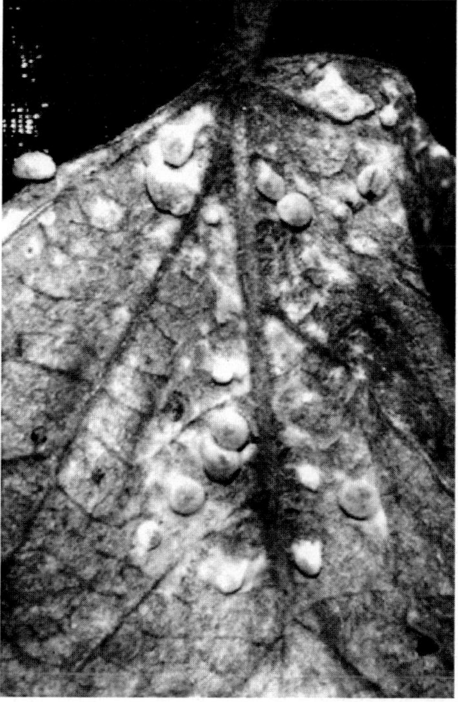

Figure 6.1 Rhizoctonia Blight of Soybean.

Casual Organism and Etiology

Rhizoctonia blight of soybean is caused by *Rhizoctonia solani* (Teleomorph: *Thanatephorus cucumeris*). Mycelium is septate, hyaline, pale to dark-brown with branching at nearly right angles in older hyphae, constriction at the point of origin of the branch. Micro- and macrosclerotia are irregular or oval, nearly uniform texture and varying in size and shape, white when young, and light to dark brown at maturity.

Disease Cycle and Epidemiology

The fungus *Rhizoctonia solani* survives on plant residue or in the soil as sclerotia and serves as primary source of inoculum. *Rhizoctonia solani* isolates generally do not produce vegetative or asexual spores, and the role of basidiospores as an inoculum source for the seedling diseases they incite on soybean is unknown. Sclerotia germinate to form mycelia and serve as an inoculum source for infection and disease spread. Continuous free moisture and a temperature 25°C are required for 24–48 hours for germination of sclerotia. Mycelia of the fungus may grow up the stem of the plants or the sclerotia or mycelial fragments may be splashed by rain onto the foliage. The disease is more severe on light and sandy soils. During warm, wet weather, mycelium spreads extensively on the surface of plants, forming localised mats of "webbed" foliage. Infection usually begins at flowering during prolonged periods of high humidity and warm temperatures. When plants mature with a closed canopy across the field, narrow rows, lodged plants and frequent rains occur, aerial blight spreads very fast. Small fields bordered by trees or poorly drained fields are more suitable to have severe aerial blight. The pathogen spreads mainly through contact between infected and non-infected parts of the plant by the formation of mycelial bridges.

Management

Cultural Control

- Avoid dense planting.
- Completely cover plant residue by clean ploughing the field soon after harvest.
- Destroy infected stubble.
- Deep summer plowing.
- Improved drainage help in reducing the disease.
- Wide row spacing or lower plant populations help in the reduction of aerial blight.

Host Resistance

- Use resistant/tolerant varieties: PK 472, PK 564, PK 1042, and SL 295.

Chemical Control

- Treat the seeds with Thiram + Carbendazim (2:1) @ 2.5–3.0 g/kg of seed. Seed treatment limits disease development in early season.
- Foliar spray of Carbendazim @ 1 g/L of water, 45–60 days after sowing.

6.2 SEED AND SEEDLING ROTS

Seed and seedling diseases of soybean are common and significant problems. Soybean seed and seedling diseases are caused by a variety of soilborne fungi and oomycetes. They can kill and rot seeds before germination or cause seedling death, reduce plant populations that result in replanting and ultimately cause huge crop losses. They are most common in heavy, compact, and poorly-drained soils or in the field with high crop residues.

Symptoms

Several different pathogens can cause these diseases, and the most common tend to be *Fusarium*, *Rhizoctonia*, *Phytophthora*, and *Pythium*. In some situations and locations, other soilborne or seedborne fungi such as *Cercospora sojina* (frog-eye leaf spot) and *Cercospora kikuchii* (Cercospora leaf blight), *Macrophomina phaseolina* (Charcoal rot), can attack seed and seedlings of soybean, causing pre- or post-emergence damping off or both. Symptoms of seed and seedling diseases may not always be typical. General and specific symptoms of seed and seedling rots caused by different pathogen are given in Tables 6.1 and 6.2.

Table 6.1 General symptoms of seed and seedling diseases of soybean caused by different pathogens

Disease/growth stage	Pathogen	Symptoms
Seed rot	*Pythium Phytophthora Phomopsis, Rhizoctonia, Fusarium*	Soft decay of seed; missing seedlings in a row.
Root and lower stem decay	*Rhizoctonia, Fusarium, Phytophthora*	Reddish-brown lesions on taproot and hypocotyl; often superficial. *Phytophthora* causes brown lesions on the stem above soil-line.
Seedling mortality (damping off, seedling blight)	*Phytophthora, Rhizoctonia, Pythium*	Wilting, yellowing (chlorosis) of leaves, necrotic lesions on stems, quick death of seedlings, leaves remain attached to stem.

Table 6.2 Comparison in specific symptoms of seed and seedlings rot of soybean

Seedling diseases	Diagnostic symptoms	Field patterns
True Fungi		
Rhizoctonia root rot and seedling blight	**Seed symptoms:** Seed rot **Seedling symptoms:** Sunken and dry reddish-brown lesions on the hypocotyl, often have a canker-like appearance usually at the soil line (Figure 6.2).	• Patchy or entire row sections • Severe in sandy soils
Fusarium root rot	**Seed symptoms:** Usually does not cause seed rot. **Seedling symptoms:** Dry, light- to dark-brown lesions on roots with chlorotic and scorch leaves, promote adventitious root growth near the soil line.	• Patchy
Phomopsis seed and seedling rot	**Seed symptoms:** Seeds are cracked and shriveled; often covered with chalky, white mold. **Seedling symptoms:** Reddish-brown, pinpoint lesions on the cotyledons or reddish-brown streaks on the stem near the soil line.	• Patchy
Oomycetes		
Phytophthora root rot	**Seed symptom:** Seeds rot **Seedling symptoms:** Soft and water-soaked, chocolate brown lesions on stem and roots; leaf chlorosis, necrosis, stunting, and plant wilting; pre- and post- emergence damping off.	• Patchy • Most common in low-lying or compacted areas prone to flooding
Pythium seedling blight	**Seed symptoms:** Soft, mushy, slimy rot of seeds in the ground. **Seedling symptoms:** Soft, brownish-coloured, rotting tissue, water-soaked lesions on the hypocotyls or cotyledons; damping off; seedlings may easily be pulled from the soil because of rotted roots; older plants yellow, stunted, or wilted if the infection is severe.	• Patchy • Most common in low-lying or compacted areas prone to flooding

Casual Organism and Etiology

- Rhizoctonia root rot and seedling blight: *Rhizoctonia solani*
- Fusarium root rot: *Fusarium* spp. (*Fusarium solani*, *F. oxysporum*, *F. tricinctum*, and other *Fusarium* species)

- Phomopsis seed and Seedling rot: *Phomopsis longicolla.*
- Phytophthora root rot: *Phytophthora sojae*
- Pythium seedling blight: *Pythium* spp. (*Pythium ultimum* and other *Pythium* species)

Figure 6.2 Soybean root rot caused by *Rhizoctonia solani.* Rusty-brown lesions on stems *Courtesy:* Dr. Dean Malvick, Department of Plant Pathology, University of Minnesota, USA.

Disease Cycle and Epidemiology

Rhizoctonia solani

The fungus survives on plant residue or in soils as sclerotia. The disease is more common in wet soils or moderately wet soils where germination is slow or emergence is delayed. Normally appears as the weather becomes warm (27°C); more often seen in late-planted soybean fields. When soils warm, the fungus becomes active and infection may occur soon after the seed is planted. The fungus grows better in aerated soils; thus, the disease is more severe on light and sandy soils. Symptoms may disappear if infected plants grow out of the root rot problems although plants may remain stunted.

Fusarium spp.

The fungus survives in the soil either as spores or as mycelium in plant residue. Certain weeds may serve as hosts to some pathogenic *Fusarium* species. The fungi can infect plants at any stage of soybean development but infection is particularly favoured when plants are weakened. Stresses such as herbicide injury, high soil pH, iron chlorosis, nematode feeding and nutritional disorders can all predispose plants to infection. After infection, damage to plants can be worsened if soil moisture is limited because of the compromised root systems. Some *Fusarium* species thrive in dry conditions, but many *Fusarium* species that cause seedling diseases are favoured cooler and wet conditions.

Phomopsis longicolla

The fungi survive winter in infected seed, infested crop residue, and weeds. Infection can occur early in the growing season without causing symptoms. Disease is favoured by warm, humid weather when soybean plants are maturing. Seed and seedling decay caused by the fungal complex *Phomopsis longicolla* and *Diaporthe phaeseolorum* var *sojae* is very common.

Phytophthora sojae

Phytophthora sojae survives on crop residue or in the soil as oospores. Optimum soil moisture is 15% to 20% is needed for oospores germination to produce structures that release swimming spores, called zoospores, under saturated soil conditions. The zoospores are attracted to soybean roots. Infection occurs via the roots, and from there the pathogen colonizes the roots and stems. Disease is most common in poorly drained soils, but may occur in other soils as well. *Phytophthora sojae* generally grow rapidly in warm, wet soil conditions. Germination of oospore occurs when soil is saturated and soil temperature exceeds 16°C, and hyphae from oospores can directly infect plants or form sporangia. The sporangia produce zoospores that are mobile in water for short distances, and can further move in water movement through soil. Zoospores are attracted to seed and root exudates, causing the spores to germinate and penetrate plant roots.

Pythium spp.

The pathogen survives either in plant residue or in soil as oospores. Oospores serve as the primary source of inoculum. Infection of soybean can occur soon after planting when hyphae or zoospores produced by sporangia reach the seed or root. As *Phytophthora*, *Pythium* produces zoospores that swim in free water and infect the seeds or roots of plants. Reproduction in infected tissue may happen rapidly but secondary infections are limited. In general, *Pythium* species are prevalent at lower temperatures (10°C–15°C), but some *Pythium* species are at warmer temperatures (30°C–35°C). The severity of the disease depends on the amount of the pathogen in the soil, plant age, and environmental conditions at the time of infection. Saturated soil is critical for infection for all *Pythium* species.

Management

Cultural Methods

- Plant certified seed.
- Remove crop residues and weeds.
- Follow crop rotation for long periods. However, seedling disease pathogens have a very broad host range which limits suitable non-host crops for rotation.
- Crop rotation and tillage will reduce the survival of *Diaporthe* and *Phomopsis* species. Use non-host crops for crop rotation including corn.
- Avoid planting into cold, wet soils, especially during zero-till operations.
- Avoid planting soybean in poorly-drained fields and compacted soil.
- Maintain proper soil fertility throughout the growing season.
- Ridging soil around stems stimulates new roots.
- Avoid damage to developing seedlings from herbicides, insects, or other physical factors.

Host Resistance

- Grow resistant varieties of soybean.

Chemical Control

- Use broad-spectrum fungicides for seed treatment such as Bavistin or Benlate (2 g/kg) are effective against *Rhizoctonia* and the same fungicide can be used as foliar sprays 2–3 times gives good control.
- If *Fusarium* is a problem in a field, seed treatments with Bavistin @ 2 g/kg seed may protect seedlings in subsequent years.
- Where *Phytophthora sojae* is a serious problem, seed treatments with Metalaxyl or Ridomil MZ72 @ 2 g/kg of seed.
- Fungicidal seed treatments with Thiram, Ziram, and Apron are effective against *Phomopsis* species.
- Where *Pythium* is a problem, treat the seed with Apron, Metalaxyl, or Strobilurins. Soil application of Metalxyl at transplanting time followed by weekly sprays of potassium phosphonate (1.0 g/L) + Acibenzolar-S-Methyl (0.025 g/L) to reduce root rot infection.
- Strobilurin fungicide (Azoxystrobin, Trifloxystrobin, or Pyraclostrobin) may help to reduce damage from true fungi such as *Fusarium* and *Rhizoctonia.*
- Treat the seed with Mefenoxam or Metalayl can be effective against *Pythium* and *Phytophthora*, and products containing Fludioxonil.

6.3 BACTERIAL DISEASES

6.3.1 Bacterial Pustules

The bacterial pustule of soybean was first described from Japan by Nakano (1919). In a study reporting 11% loss in yield, 63% of it was due to the reduction in seed size. In India, the disease was first reported by Uppal et al. (1938) from Jalgaon (Maharashtra). Now, it is present in most parts of the country. Severe infection causes premature defoliation that decreases yield by reducing size and number of seeds. The diagnostic symptom of bacterial pustule is the presence of pustules, which are minute, pale-green spots with elevated centres, usually on the abaxial surface of leaves, formed through hypertrophy and hyperplasia. When foliage is covered with pustules, premature defoliation can occur, reducing yields by reducing seed size and quantity.

Symptoms

The symptoms are confined mostly to the leaves and are most prevalent during the later growing season of the crop. The symptoms start as small, pale, yellowish-green spots (lesions) having dark reddish-brown centers, formed mostly on the upper leaf surface (Figure 6.3). Lesions are often associated with main leaf veins. As the disease progresses, the central part of each lesion develops into a minute, raised, light-coloured pustule usually on the lower surface of the leaf. Sometimes, leaf spots are formed without developing pustules. The lesions merge with each other forming large, irregular areas of dead leaf tissue surrounded by yellowish margins. The leaves become ragged when parts of the brown, dead areas tear away during windy, rainy weather. Severe infection often results in partial defoliation. On the pods, small, reddish brown, slightly raised spots may develop on the susceptible cultivars. The symptoms of soybean pustule can be confused with those of soybean rust (Table 6.3).

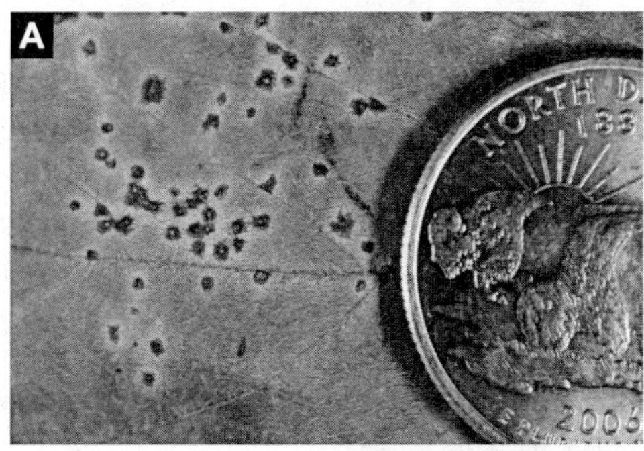

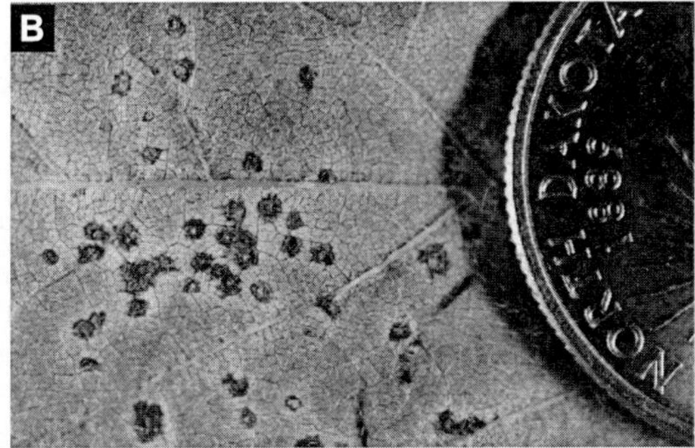

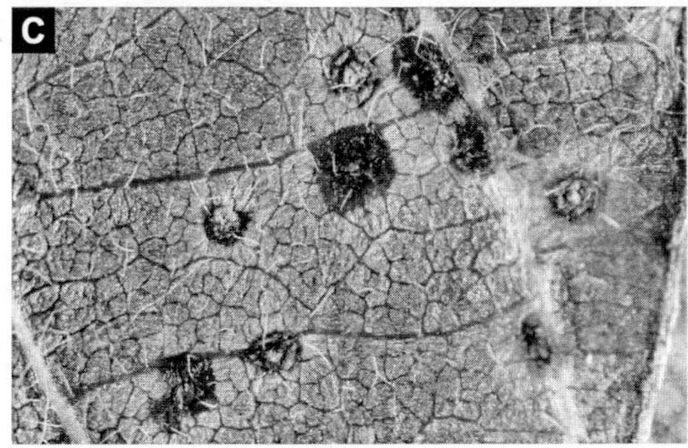

Figure 6.3 Bacterial Pustules of Soybean. (A) Pustules on upper surface, (B) Pustules on lower surface, (C) Close up of bacterial pustules. *Courtesy:* Dr. Samuel Markell, Department of Plant Pathology, North Dakota State University, USA.

Table 6.3 Difference in rust pustules and bacterial pustules of soybean

Rust pustules	Bacterial pustules
Mature rust pustules have a round opening on the upper leaf surface for the release of spores.	If any opening is present, it is typically a linear crack across the surface of pustule.

Note: These features can only be seen with magnification of 20X.

Causal Organism and Epidemiology

Bacterial pustule of soybean is caused by *Xanthomonas axonopodis* pv. *glycines* (Syn. *X. glycines*, *X. campestris* pv. *glycines*). The bacterium being a pathovar of *X. axonopodis* species has the same morphological, cultural, and biochemical characters as those of *X. axonopodis*. It can be distinguished from *X. axonopodis* and its other pathovars on the basis of host range. The description of *X. axonopodis* is given under bacterial leaf spot of green gram. Colonies on leaf infusion agar are small, circular, smooth having entire margin; pale yellow, but become deep yellow with age. The bacterium grows slowly in culture. The optimum growth occurs at 30°C–33°C. The maximum and minimum temperatures at which growth can take place are 38°C and 10°C, respectively.

Disease Cycle and Epidemiology

The bacterium survives in the infected crop residue, infected seed, and the rhizosphere of wheat roots. The bacterium can survive in the seed for more than 2 years. The chances of survival are more in infected crop residue when it remains on the soil surface. The entry of the bacterium occurs through stomata and wounds. After entry into the host, the bacterium moves into intercellular spaces and multiplies. In response to infection, leaf cells increase in size and also divide at a faster rate. Increase in host growth and in bacterial cell mass in the localised area causes bulging of epidermal cells on both leaf surfaces. These raised areas rupture and become pustules. The dissemination of the bacterium occurs through wind-splashed rain, overhead-irrigation, and cultural operations during wet foliage. Disease outbreaks occur after several days of rainstorms or hailstorms. The disease is favoured by warm and wet weather. The optimum temperature for disease development is 30°C–33°C.

Management

Chemical Methods

- Use pathogen-free seed.
- Follow crop rotation with non-host crops for 2–3 years.
- Destruction of infected crop residue by deep ploughing.
- Avoid dense cropping.
- Avoid or limit the use of overhead-irrigation.
- Avoid cultural operations when foliage is wet.

Host Resistance

- Use of resistant varieties is the most effective method for the management of this disease. Grow resistant or tolerant varieties such as PK 1029, PK 1042, Indira soya 9, JS 95-60, JS 97-52, JS 335, JS 93-05, Bragg, KHSB 2, MAUS 32, NRC 7, NRC 37 and VLS 2.

Chemical Control

- Treat the seed with Streptocycline @ 250 ppm (2.5 g per 10 kg seeds).
- Spray Copper oxychloride @ 3 g/L of water.

6.3.2 Bacterial Blight

Bacterial blight of soybean was first noticed by Heald in Nebraska, USA in 1905. It is the most common bacterial disease of soybean in areas where cool and wet weather prevails. Bacterial blight of soybean cause yield reduction upto 17.9%. The bacterium also considerably inhibits the germination of infected seeds and inhibition up to 68%. The average disease incidence upto 14.5% and maximum pod blight severity have been recorded in Marathwada region of Maharashtra (India) during the Kharif seasons of 2009 and 2010.

Symptoms

The symptoms appear on all the aboveground plant parts but are most evident on leaves in the mid to upper canopy. Generally, the lesions first appear on cotyledons as brown spots on margins, which enlarge, and turn dark brown as the tissue collapses. Young plants may show stunting and may die if the infection reaches the growing point. Symptoms are most conspicuous on leaves although they also appear on stems, petioles, and pods. Young leaves are most susceptible. Symptom on the leaves starts as small, angular, yellow to brown lesions. The lesions are usually water-soaked in the center and surrounded by yellowish-green haloes (Figure 6.4). Older lesions turn reddish-brown to black as the tissue dies. Angular lesions enlarge in cool and wet weather and may coalesce to produce large, irregular dead areas. Eventually the dried lesions fall out giving the leaves a ragged appearance. Often, the leaves are badly shredded after strong winds and stormy rains. Infected young leaves become stunted, distorted, and chlorotic. Severe infection results in defoliation of lower leaves. Large, black lesions may also develop on stem and petioles. The symptoms on seed appear rarely. However, when lesions appear on pods, developing seeds may show some kind of symptoms. Small, water-soaked lesions appear on pods, which enlarge and merge giving rise to dark brown to black areas. Infection/infestation of seeds occurs through the pods during the growing season or at the time of threshing and cleaning. The slimy bacterial ooze may cover the developing seeds. During storage, infected/infested seeds may become shriveled, discoloured, and develop sunken or raised lesions. Infected seeds lack vigor and germinate poorly. Chlorosis, also called "systemic toxemia", develops throughout the infected plants. It results from the production of a toxin, coronatine, by the bacterium in the older infected leaves and translocated throughout the plant. Toxin interferes with photosynthesis and chlorophyll production leading to systemic chlorosis and stunting of plants. Fully expanded trifoliate leaves do not show systemic chlorosis, while it is maximum on young trifoliate leaves. The symptoms of bacterial blight can be confused with those of soybean bacterial pustules. See the difference between these two bacterial diseases in Table 6.4.

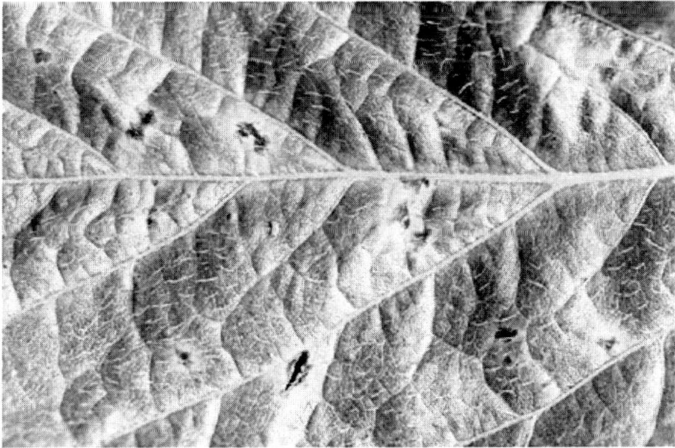

Figure 6.4 Bacterial Blight of Soybean. *Courtesy:* Dr. Samuel Markell, Department of Plant Pathology, North Dakota State University, USA.

Table 6.4 Difference in bacterial blight and bacterial pustules of soybean

Features	Bacterial blight	Bacterial pustules
Causal organism	*Pseudomonas savastanoi* pv. *glycinea*	*Xanthomonas axonopodis* pv. *glycines*
Initial symptoms	Yellow-brown spots with water soaked leaf lesion.	Tiny pale green spots on leaves.
Later symptoms	Spots are flat or more angular in shape with water-soaked tissue and yellow halo.	Circular spots with raised centers and water-soaking is absent.
Seasonal development	Early season	Late season
Survival	Infected crop residues and seed.	Infected crop residues and seed.
Favorable environmental condition	Cool, wet weather and temperature 21°C–27°C.	Warm, wet weather and temperature 30°C–33°C.
No disease development	Below 4°C and above 35°C.	Not checked by high temperatures.

Disease Cycle and Epidemiology

The bacterium *Pseudomonas savastanoi* pv. *glycinea* overwinter in crop residue, infected/infested seeds, and volunteer bean plants. Therefore, bacterial blight is favoured by continuous soybean cropping, zero-tillage production systems, or fields planted with infected seed lots. The bacterium grows epiphytically on the leaves, buds, and stems of soybean plants without causing any symptoms. This epiphytic phase may serve as a continuous source of inoculum. It is typically an early-season disease favoured by cool and wet weather, which prevails during the early crop season. The optimum temperature for disease development is 21°C–27°C. Warm and dry weather after the midseason of the crop prevents the disease development. Disease epiphytotics are commonly seen after windy rainstorms coupled with cool weather. Primary infection usually occurs on the cotyledons, and these lesions serve as a major source of secondary inoculum. The spread of the bacterium occurs through wind-splashed rain, overhead-irrigation, and cultural operations especially during wet foliage and rubbing of wet leaves with each other. After entering through stomata and wounds, bacteria multiply in the intercellular spaces of mesophyll cells. Typical water-soaked lesions develop 5–7 days after infection. During multiplication in the host, pathogen produces toxin called coronatine, which is translocated throughout the plant, mainly through the phloem.

Management

Cultural Methods

- Use pathogen-free seed for sowing.
- Rotate the crop with maize and sorghum but not with lima beans or snap beans.
- Bury residue or place it in closer contact with the soil for faster decomposition.
- Follow crop rotation with non-host crops for 2–3 years
- Avoid the cultural operations during wet surface of foliage.
- Avoid the overhead-irrigation
- Practice deep tillage in fields where the disease is frequently severe.

Host Resistance

- Grow resistant varieties.

Biological Control

- Use bioagent such as *Bdellovibrio bacteriovorus* which inhibit local and systemic disease development of bacterial blight.

Chemical Control

- Spray Streptocycline (@ 100 μg m/L) + Copper oxychloride (@ 0.25%).

6.3.3 Wildfire

Wildfire is one of the most destructive bacterial diseases and was recognised as a disease of soybean in 1943.

Symptoms

Light brown to black, dead spots that are 2–15 mm or more in diameter and of variable shape develop on the leaves. These lesions are nearly always surrounded by a broad conspicuous yellow halo. The yellow halo distinguishes wildfire from other bacterial diseases of soybean. In damp weather, the lesions enlarge and merge forming large, dead areas in the leaf that become dry, tear away, and produce tattered leaves. A bacterial pustule is almost invariably present in the center of a wildfire lesion.

Causal Organism and Etiology

Wildfire disease is caused by *Pseudomonas syringae* subsp. *tabaci* (synonym: *P. tabaci*). It is a hemibiotrophic bacterial pathogen. The bacterium produces a fluorescent pigment and a potent toxin, called tabtoxin or wildfire toxin. A mere 0.05 milligrams of this toxin can produce a yellow lesion on a tobacco leaf in the absence of bacteria. The bacterium produces a hypersensitive reaction when injected into the leaves of tomato and pepper. Tabtoxin is produced by the bacterium *Pseudomonas syringae*; pv. *tabaci*, which causes the wildfire disease of tobacco; by strains of pv. *tabaci* occurring on other hosts such as bean and soybean; and by other pathovars (subspecies) of *P. syringae*, such as those occurring on oats, maize, and coffee. Toxin-producing strains cause necrotic spots on leaves, with each spot surrounded by a yellow halo.

Disease Cycle and Epidemiology

The wildfire bacterium overwinters in infected crop residue and seeds. The bacterium multiplies on the root surfaces of many crop and weed plants. The bacterium is spread by splashing water and wind-blown rain. Free water is often required for the invasion and infection of soybean leaves. The optimum growth of the bacterium occurs at temperatures 24°C to 28°C.

Management

Cultural Methods

- Use pathogen-free seeds.

- Remove and destroy diseased plants.
- Rotate soybeans for one year or more with corn, sorghum, alfalfa, clovers, or cereals.
- After harvesting, completely cover the stubble and plant residue by deep ploughing.
- Avoid the cultural operations during wet surface of foliage.

Host Resistance

- Grow resistant varieties like Ogden and Lee.

6.4 MOSAIC DISEASES

6.4.1 Soybean Mosaic

It is one of the most common viral pathogens of soybean. It can cause significant damage to soybean and is present worldwide. Infection by SMV usually results in severe yield losses (8 to 50%) and seed quality reduction. Yield losses as high as 86% have been reported in uniformly infected field plots. The extent of damage to the crop is dependent on the host genotype, predominant virus strain, infection incidence, and the developmental stage at which soybean plants become infected.

Symptoms

Symptoms vary with host genotype, virus strain, plant age at infection, and environment. The commonly observed symptoms include moderate to severe leaf mottling, mosaic, rugosity, dark green vein banding, and light green interveinal areas, overall stunting, leaf curling distortion of leaves, browning of stems and petioles male sterility, flower deformation, seed coat mottling, bleeding hilum, seed discolouration, necrosis, sometimes necrotic local lesions, systemic necrosis, bud blight, and occasionally death of the infected plants (Figure 6.5).

Figure 6.5 Soybean Mosaic. (A) Severe Leaf mottling and curling, (B) Seed discoloration.
Courtesy: Albert Tenuta, Ontario Ministry of Agriculture, Food and Rural Affairs (OMAFRA), Ridgetown, Ontario, USA.

Causal Organism and Etiology

Soybean mosaic is caused by *Soybean mosaic virus* (SMV). It is a member of the genus Potyvirus in the family Potyviridae. Its virion consists of nonenveloped, filamentous, flexuous rod-shaped particles with 650–760 × 15–18 nm in dimension. The genome of SMV is a monopartite positive ssRNA molecule.

Host range: *Soybean mosaic virus* has a very narrow host range which is limited to six plant families: Fabaceae, Amaranthaceae, Chenopodiaceae, Passifloraceae, Schrophulariaceae, and Solanaceae. The most commonly infected hosts are wild soybean (*Glycine soja*) and cultivated soybean (*Glycine max*).

Disease Cycle and Epidemiology

Soybean mosaic virus is transmitted by aphids, sap, and by infected seeds. About 32 different species of aphid transmit SMV, including *Acyrthosiphon pisum*, *Aphis fabae*, *Myzus persicae*, *Aphis craccivora*, *Macrosiphum euphorbiae*, *Rhopalosiphum maidis* and *R. padi*, in a nonpersistent manner. Aphids become viruliferous by feeding on infected stems, stem tips, and old or young leaves. Virus may persist for up to two years in the seed. About 30% or more of the seeds from SMV-infected soybean plants carry SMV depending on the cultivar and time of infection before flowering. Weeds and other plants may also serve as a reservoir for SMV. Early plant infection reduces pod set, increases seed coat mottling and reduces seed size and weight. Plants that become infected by seed transmission serve as primary inoculum sources for SMV. The viruses must multiply in infected cells for translocation and systemic infection. SMV moves up and down in plants and can be detected in all parts of systematically infected plants. Disease is severe at 26°C and above.

Management

Cultural Methods

- Use of SMV-free seeds for planting is an effective way of preventing the disease.
- Remove virus-infected plants and weed hosts from fields and burn them.
- Do not delay seed sowing. Late planting coincides with higher populations of aphids, which may increase the probability of virus transmission to young seedlings.
- Maintain a dense soybean canopy.
- Since SMV is sap-transmitted, avoid handling or brushing up against infected plants.

Host Resistance

- Use resistant or tolerant varieties; JS 71-05, KHSB2, LSB 1, MACS 58, MACS 124, Punjab 1 and VLS2.

Chemical Control

- Spray insecticides (Pyrethroid or Organophosphate) to suppress soybean aphid populations.

6.4.2 Yellow Mosaic

Yellow mosaic diseases (YMD) are major constraints in the productivity of grain legumes in India. Yield loss per annum due to YMD has been estimated to be $300 million including black gram, green gram, and soybean crops together. Yellow mosaic disease of soybean was first described by Nariani (1960). The yellow mosaic disease causes 15%–75% yield loss in soybean production.

Symptoms

Typical symptoms like yellow mosaic, mottling and crinkling on the leaf, and stunted growth of the plants. The most prominent symptom on the foliage starts as small yellow specks along the vein-lets and spreads over the lamina; the pods turn thin and curl upwards. Rusty necrotic spots appear in the yellow areas as the leaves mature. Under severe condition, plants produce shriveled and lightweight seeds or sometimes fail to form flowers and pods.

Casual Agent and Etiology

Yellow mosaic disease of soybean is caused by *Mungbean yellow mosaic virus* (MYMV). MYMV belong to the genus, Begomovirus of the family, Geminiviridae. Begomoviruses have characteristic icosahedral geminate particles that encapsidate genome of circular single-stranded DNA. They have monopartite or bipartite genome. In bipartite begomoviruses, DNA A encodes proteins required for replication, transcription and encapsidation whereas DNA B encodes proteins required for movement functions.

Host range: Mungbean, black gram, kidney bean, azuki bean, *Alternanthera sessilis*, *Sida rhombifolia*, etc.

Disease Cycle and Epidemiology

MYMV is transmitted by the whitefly (*Bemisia tabaci*) in a persistent (circulative) manner and grafting but not by sap and seed. The disease agent has been transmitted experimentally by whiteflies, and symptoms developed after 23 days. Symptoms are more severe at lower temperatures.

Management

Cultural Methods

- Use pathogen-free seed.
- Removal of crop residues and weeds is effective against whiteflies.
- Deep summer ploughing.
- Remove virus infected plants and weeds hosts from fields and burn them.
- Avoid excess use of nitrogen fertilizer, including manures, as succulent growth will increase whitefly populations.

Host Resistance

- Use resistant or tolerant varictics; JS 97-52, JS 20-29, JS 95-60, PK 472, PK 564, PK 1024, PK 1029, PK 416, PK 1042, Pusa 37, SL 295, SL 525, and SL 688.

Chemical Control

- Treat seed with Thiamethoxam (70 WS) @ 3 g/kg seed.
- Spray insecticide to reduce insect vector population, i.e. Metasystox 25 EC @ 1 ml or Imidachloprid @ 0.5 ml/L of water or Thiamethoxam (25 WG) @ 100 g/ha, after 35 days of sowing.

7 Pigeon Pea (*Cajanus cajan* [L.] Millsp.)

7.1 PHYTOPHTHORA BLIGHT

Phytophthora blight is the third potentially important disease of pigeon pea after Fusarium wilt (*Fusarium udum*) and pigeon pea sterility mosaic disease (*Pigeon pea sterility mosaic virus*). Phytophthora blight is a potentially emerging threat to pigeon pea production and can cause 100% yield losses under a favourable environment. The first suspected occurence of Phytophthora blight on pigeon pea in India was reported by Williams et al. in 1966.

Symptoms

The symptoms of the Phytophthora blight on pigeon pea have been described as stem rot, stem blight, stem canker and root rot. Based on the symptoms, the disease has been designated as stem rot, stem blight, stem canker, and Phytophthora blight.

Foliar Blight

Characteristic foliage blight symptoms first appear as water-soaked lesions on the primary and trifoliate leaves which become necrotic within a week. The leaflet lesions are circular to irregular in shape and can be upto one centimeter in diameter. Under conditions of high humidity, the whole foliage gives a blighted appearance.

Stem Rot, Stem Blight, and Stem Canker

Stem symptoms appear as brown to dark brown lesions distinctly different from healthy green portions on the main stem, branches and petioles. The lesions on stems and branches increase rapidly and extend up to 20 cm, girdle and cracks and dry the stem. Conversion of stems swelling into cankerous structures is very common. Sometimes lesion areas crack and shred.

Infected stems and branches break easily by wind. Phloem vessels show smoky grey-coloured discolouration and xylem vessels remain healthy. As the disease progresses, patches of diseased plants become conspicuous in the field which is visible even from a distance. General symptoms of Phytophthora blight and Fusarium wilt of pigeon pea appear similar and cause confusion in diagnosis. See Table 7.1 for the difference between these two diseases.

Table 7.1 Difference in symptoms of Phytophthora blight and Fusarium wilt in pigeon pea

Plant part/ growth stage	Phytophthora blight	Fusarium wilt
Young seedling	Killed within 3–10 days.	Gradually wilt and die within 10–30 days; retain the dull green colour.
Foliage	Water-soaked lesions on the leaves and whole foliage gives a desiccated appearance.	Loss of turgidity, slight chlorosis, and drooping of leaves.
Stem	Distinct brown to dark brown lesions, girdling of stem; phloem smoky grey coloured and clear xylem; swelling at the base or cankerous hypertropical structures; wind-break easily.	Dark brown to purple streak band usually seen only on one side of the stem; browning of the xylem vessels; pinkish mycelial growth on the part of the wilted plants.
Root	Healthy and cannot be uprooted easily.	Dried and can be pulled easily.

Causal Organism and Etiology

Phytophthora drechsleri f. sp. *cajani* causes stem blight of pigeon pea. Hyphae are hyaline, coenocytic devoid of granular contents, and slender. Sporangiophores are hypha-like. Sporangia are ovate to pyriform, rarely spherical with minute papilla. Zoospores are biflagellate, hyaline ovoid to reniform, tapering slightly at the anterior end. Antheridia are simple, hyaline, amphigynous, and persistent. Oospores are spherical to globose. Chlamydospores are intercalary and terminal.

Disease Cycle and Epidemiology

Phytophthora drechsleri f. sp. *cajani* is capable of surviving in soil and infected crop debris in the form of oospores and chlamydospores for at least one year. At the onset of the rainy season, oospores germinate by producing sporangia (zoospores) and direct germ tube and incite infection in young seedlings. Zoospores are the primary source of inoculum and that wind contributes to inoculum dispersal over short distances during rains. From primary infection, a large number of sporangia are produced and serve as secondary inoculum which is disseminated by wind, water, movement of soil, and raindrop splashes. High humidity (more than 80%) and a temperature range 28°C–32°C are favourable for rapid disease development in the field. The absence of potassium (K) and high doses of nitrogen (N) increase Phytophthora blight incidence.

Management

Cultural Methods

- Use pathogen-free seeds for planting.
- Select the fields with no previous history of Phytophthora blight occurrence.

- Avoid sowing pigeon pea in fields with low-lying patches that are prone to water-logging.
- Sow pigeon pea on ridges and provide good drainage.
- Use wide inter-row spacing.
- Do intercropping with short leguminous crops such as mungbean and urdbean.
- Rotate crop with non-host crops such as cereals.
- Remove and destroy alternative host species such as *Atylosia* spp. and wild *Cajanus* spp.
- Do plastic mulching before sowing to reduce soilborne inoculum.
- Use potassium-based fertilizers @ 25–50 kg/ha.

Host Resistance

- Grow variety with multiple resistances. Genotypes IPAC 3-2, IPAC-42, IPAC-79, IPAC 66-7, IPAC 66-9, IPAPB 7-2-1-7, and WD 5-4 are resistant to Phytophthora stem blight.

Biological Control

- Use bioagents such as *Trichoderma viride*, *T. hematum*, *Pseudomonas fluorescens*, and *Bacillus subtilis* which are antagonist to *Phytophthora drechsleri* f. sp. *cajani* and these are also compatible with different levels of fungicides such as Ridomil, Captan, Captafol, Thiram, and Carbendazim.

Chemical Control

- Treat the seeds with Ridomil MZ 72 (Metalaxyl + Mancozeb) @ 3 g/kg of seed.

7.2 WILT

Vascular wilt of pigeon pea is common throughout India and is considered the most destructive fungal disease of this important pulse crop. It was first recorded by Butler (1906) in India. The incidence of the disease has been reported from 30% to 60% at flowering and crop maturity stages, however, it can also cause yield losses of upto 100% in susceptible cultivars. Wilt disease can cause up to 50% plant mortality when the crop is continuously grown in the same field. The damage by this disease is considerably enhanced, if the plants are infested by cyst nematode (*Hetrerodera cajani*) and root-knot nematode (*Meloidogyne* spp.).

Symptoms

The symptoms can be observed at the flowering or pod formation stage. Wilting caused by *Fusarium udum* f. sp. *cajani* is characterised by dropping, loss of turgidity, slight interveinal chlorosis, internal browning of xylem vessels, and a purple band on stem extending upwards from the base, gradual and sometimes sudden yellowing, withering and drying of leaves and subsequently whole plant or some of its branches (Figure 7.1). The disease can be confirmed by the browning of the xylem vessels. Wilting can be partial or total. Partial wilting is mainly associated with lateral root infection, whereas tap root infection may cause complete wilting.

Recovery of the affected plants is rare. White to pinkish cottony aerial fungal growth may also occur on the base of the stem near soil-line. Random patches of wilted plants can be seen throughout the field.

Figure 7.1 Wilt of Pigeon pea. (A) Drooping of leaves (initial stage), (B) Yellowing and drooping of leaves (later stage).

Causal Organism and Epidemiology

Vascular wilt of pigeon pea is caused by *Fusarium udum* (teleomorph: *Gibberella indica*). The mycelium is hyaline, highly branched and produces 3 types of spores viz. macroconidia, microconidia, and chlamydospore. Microconidia are small, elliptical or curved, and unicellular. They may produce microconidia freely on hyphal branches or clustered together in a ball or false head in culture medium. Macroconidia are produced on small sporodochia. Macroconidia are long, curved (fusaroid), with prominent apical hook and notch at the base and septate (3–4 septa). Chlamydospores are oval or spherical, single or in chains, terminal or intercalary and persist in soil for a long period. The perfect stage, *Gibberella indica* produces superficial, commonly clustered, sub-globose to globose, sessile, smooth walled, and dark voilet perithecia on exposed roots and colour region of the stem. Ascospores are ellipsoidal to ovate, commonly hyaline 2-celled, rarely 3–4 celled, and constricted at septa.

Disease Cycle and Epidemiology

The wilt pathogen, *F. udum* survives in soil, plant stubbles, and on pigeon pea seed. However, major infection occurs through the soil and plays a major role in the transmission of the *F. udum* propagules. Dissemination also occurs through seed-borne infection. In absence of the host, the fungus can survive in the soil for upto 3–10 years. The fungus can survive in plant stubbles for 2.5 years. Weeds and cultivated plants are symptomless carriers for wilt disease and could be the possible sources of primary inoculum. The spread of disease from plant to plant occurs through root contacts, air, irrigation, rainwater, flooding during rains, termites, and nematodes. Both imperfect and perfect states are important in completing the wilt disease cycle. The fungus is a facultative parasite remains in soil as a saprophyte for longer duration. It attacks the roots with germ tubes arising from micro-, macroconidia, and chlamydospores, reaching the vascular tissues to establish and multiply rapidly, causing wilting of parts or entire plant. Secondary infection by conidia produced on aerial parts is rare. Fusarium wilt is favoured by a wide range of temperatures. Temperatures range of 17°C–29°C favours the disease development. Nematodes also play an important role in the spread of the disease by injuring the plants and also affecting the plant growth. Pigeon pea infestation by cyst nematode (*Heterodera cajani*) increases the wilt incidence. Root knot nematodes (*Meloidogyne javanica* and *M. incognita*) have been found associated with the breakdown of the resistance by retarding the accumulation of Cajanol an enzyme responsible for resistance in pigeon pea cultivar.

Management

Cultural Methods

- Select a field with no previous record of wilt for at least the past three years.
- Follow crop rotation for 4–5 years. Recommended crops for rotations are tobacco, sorghum, pearl millet, cotton, and resistant pigeon pea genotypes.
- Remove plant parts with their roots and do summer deep ploughing to reduce the inoculum of the pathogen.
- Uproot the wilted plant and use them as firewood or destroy them.
- The most effective and practical solution to this problem is mixing or intercropping of pigeon pea with sorghum.
- Application of Zinc sulphate during seedling emergence @ 25 kg/ha suppresses the disease in the field. Zinc abolishes production of *Fusarium* pathogenicity factor, i.e. fusaric acid (toxin).
- Presoaking of seed in Manganese solution induces resistance against infection.

Host Resistance

- Grow variety with multiple resistances as recommended for the region. Genotypes/varieties resistant to *Fusarium udum* are Narendra Arhar-1, BDN1, BDN2, MA3, ICP 8858, ICP 8859, ICP 8863, ICP 9174, KPL 43, KPL 44 and PI 397430.

Biological Control

- Treat the seeds with *Trichoderma harzianum* or *T. viride* @ 10 g/kg of seeds.
- Treat the soil with *Trichoderma harzianum* or *T. viride* @ 200 g/acre.

Chemical Control

- Treat the seeds with Carboxin 37.5% + Thiram 37.5% DS @ 4 g/kg and treat the soil @ 200 g/acre.
- Treat the seeds with Carbendazim 50 WP + Thiram 80 WP @ 1+3 g/kg of seed.
- In case of severe disease incidence in the field, drench the soil with Carbendazim 50 WP @ 1.0 g/L of water.

Integrated Disease Management

- Field sanitation + seed treatment + intercropping/mixed cropping of sorghum and use of moderately resistant cultivars has reduced wilt incidence in pigeon pea.

7.3 STERILITY MOSAIC DISEASE

The sterility mosaic disease (SMD) is the most devastating disease of pigeon pea in India. It is a serious problem in pigeon pea producing states such as Tamil Nadu, Maharashtra, Gujarat, Punjab, Uttar Pradesh and Bihar. It was first described from Pusa (Bihar), India in 1931. This disease is also known as the *green plague*, as the infected plants remain in the vegetative state (pale green with excessive vegetative growth) with poor flowering or without flowering.

Symptoms

The disease is characterised by one or more of the following features: a bushy and pale green appearance of the plants, complete or partial cessation of flower production (sterility), mosaic or chlorotic ringspot symptoms on the leaves, excessive vegetative growth, stunting, leaf distortion and reduction in leaf size (Figure 7.2). Depending on the pigeon pea genotype and the time of virus infection, the symptoms of SMD may vary and can be grouped into three categories: (i) severe mosaic and complete sterility, (ii) mild mosaic with partial sterility, and (iii) chlorotic ring spots without any noticeable sterility. These ring spots are characterised by a green island surrounded by a chlorotic halo which indicates the localised sites of infection of the pathogen.

Figure 7.2 Pigeon pea Sterility Mosaic. (A) Mosaic symptoms on leaves, (B) and (C) Infected leaves with chlorotic ring spots symptoms.

The symptoms of SMD are often masked in older plants. However, in ratooned plants, the new leaves show clear symptoms. In partially affected plants, the number of pods is reduced and seeds are discoloured, deformed, and with reduced dry weight (about 20%). Highest disease incidence and yield losses are in ratoon and perennial pigeon pea crop. The yield losses caused by SMD also vary and depend on the genotypes of pigeon pea and the growth stage at which infection occurs (Table 7.2).

Table 7.2 Timing of infection and yield losses caused by SMD

Timing of infection	Age of plant	Symptoms	Yield loss (%)
Early infection (*Before flowering*)	Less than 45 days old	Plants show characteristic symptoms and there is the complete absence of flowers (complete sterility).	95–100
Late infections (After flowering)	More than 45 days old	Plants show only mild mosaic symptoms on only on few branches and there is partial suppression (20%–50%) of flowering (partial sterility). However, after rationing (severe pruning), new growth from such plants shows severe mosaic symptoms and complete sterility.	26–97

Causal Agent and Etiology

Pigeon pea sterility mosaic disease is caused by *Pigeon pea sterility mosaic virus* (PPSMV). It is a species of the genus *Emaravirus*, and is the causal agent of sterility mosaic disease (SMD) of pigeon pea. The virion of PPSMV is thread-like flexuous, irregularly-branched, filamentous, 8–11 nm in diameter, and of undetermined length. The particles of this virus consist of at least five to seven, negative-sense, single-stranded RNA segments as a genome and a protein with an estimated size of 32 kDa.

Disease Cycle and Epidemiology

The primary source of inoculum is diseased plants left in the field, collateral hosts, perennial pigeon pea, ratooned pigeon pea, and its wild relatives. PPSMV is neither seed nor transmitted mechanically or by plant sap. PPSMV is transmitted in a semi-persistent manner by a wingless eriophyid mite *Aceria cajani* which is very small and can be seen only under a microscope. A single eriophyid mite is enough to transmit the virus and both nymphs and adults are equally efficient transmission agents. *Aceria cajani,* is an obligate pest dependent on pigeon pea during all stages of its life cycle, and it inhabits the lower surface of leaves. It is predominantly found on the symptomatic leaves of PPSMV-infected plants and their feeding causes no mechanical damage to the host plant. The *A. cajani* mites acquired PPSMV after a minimum acquisition access period (AAP) of 15 minutes and transmitted the virus after a minimum inoculation access period (IAP) of 90 minutes. There is no latent period. However, there is no evidence of PPSMV replication within the mite, which remains infectious for about 6–13 hours, and no transovarial transmission. This mite is highly host-specific and is largely confined to pigeon pea and its wild relatives, *C. scarabaeoides* and *C. cajanifolius*. Perennial and volunteer pigeon pea and ratooned growth of harvested plants provide reservoirs of mite and virus. The dispersal of these mites is passive, mainly assisted by wind currents. Under favourable conditions, it spreads across the fields in an epidemic form. Of the different abiotic factors, such as temperature, relative humidity and rainfall, relative humidity has a significant effect on the mite. A mean temperature of about 20°C–30°C is favourable for the proliferation of the mites. However, high temperatures and heavy rainfall are unfavourable for their growth.

Management

Cultural Methods

- Rouge out the infected plants upto 40 days after sowing.
- Destroy perennial or ratooned pigeon pea plants.
- Choose a field away from perennial or ratooned pigeon pea plants.
- Destroy the infected plants at an early stage of disease development.
- Rotate crops to reduce mite vector population.

Host Resistance

- Grow resistant cultivars. Some resistant cultivars are ICPL 87, C 11, ICPL 83024, ICPL 7867, BDN 1, ICPL 366, Hy 3C, ICPL 151 (Jagriti), ICPL 81, and NPWR 15.

Chemical Control of Vector

- Spraying with Fenazaquin @ 1 ml/L, soon after the appearance of the disease, and if necessary repeat after 15 days.
- Spray acricides such as Teldon, Morestan, and Kelthane all @ 0.1% are highly effective in controlling mites.
- In case of the appearance of initial symptoms (break up of green colour) in the field, look for mites. In case of presence of mites, spray the crop with either of the chemical insecticides, e.g., Dicofol 18.5 EC (2.5 ml/L) or Oxydemeton methyl 25 EC (2.0 ml/L) or Dimethoate 30 EC (1.7 ml/L).

8 Finger Millet (*Eleusine coracana* Gaertn.)

8.1 BLAST

Blast of ragi is one of the major production constraints causing serious yield losses. It was recorded first in India by McRae (1920) from Tanjore delta of Tamil Nadu. The damage depends on the time of onset and severity of the disease. The average losses of around 28% and as high as 80% to 90% in endemic areas have been reported due to finger millet blast have been reported.

Symptoms

Blast symptoms can be observed at almost every stage of plant growth such as on seedlings, leaves, peduncles, stems, roots and fingers, depending on the stage of the crop. The characteristic symptoms on leaves are water-soaked, spindle-shaped or elliptical or diamond-shaped lesions with white or grey centers, and a chlorotic halo surrounding the lesions (Figure 8.1A, B). Under favourable environmental conditions, these spots enlarge, coalesce and give a blasted appearance to the leaf blades. Neck blast symptoms develop as elongated black lesions mostly one-two inch below the ear (Figure 8.1C). Finger blast symptom starts at the tip and proceeds toward the base of the finger, which becomes brown. Neck infection is the most serious and damaging phase of the disease. The neck region turns black and shrinks. Under humid conditions, oliveceous-grey fungal growth of the pathogen may be seen in the neck area. Neck blast causes major loss in grain number, grain weight, and increase in spikelet sterility. The pathogen attacks seeds resulting in shriveled and blackened seeds.

Causal Organism and Etiology

Blast of ragi is caused by *Pyricularia grisea* (Perfect stage: *Magnaporthe grisea*). The conidia are pyriform, usually 3-celled (2 septate), and have a small hilum at the base.

Figure 8.1 Blast of Finger Millet. (A) Leaf blast, (B) Close up of a spindle shaped lesion on leaf, (C) Panicle and neck blast symptoms.

Host range: The fungus infects many other economically important crops, such as rice, foxtail millet, wheat, barley, and other grasses within the family Poaceae.

Disease Cycle and Epidemiology

Pyricularia grisea commonly survives in crop residues and on other cereals. The pathogen is also observed in the seed outside the embryo. Initial inoculum comes from infected seeds, weeds, or collateral hosts; spread by airborne conidia. One infected seed could result in a disease epidemic. *Magnaporthe oryzae* infects the host in two stages, the biotrophic stage where it obtains nutrients from live cells, and the necrotrophic stage where it obtains nutrients from dead cells. Development of the disease is favoured by a minimum temperature of 15°C–25°C and relative humidity of more than 85% with intermittent rainfall. Disease severity is influenced by various factors such as susceptibility of the genotypes, sowing time, and the corresponding environmental conditions.

Management

Cultural Methods

- Use disease-free seeds.
- Proper plant spacing and transplanting are advisable.
- Early sowing (July month) reduces the blast severity.
- Do not apply excess nitrogen to avoid neck and finger blasts.

Host Resistance

- Grow resistant varieties such as CO RA (14), Paiyur (RA)-2, GPU-26, GPU-28, GPU-45, GPU-48, VL 149, CO 13, and L-5.
- Use of varietal mixture @ 1:1 ratio of PRM 1 (susceptible) and VL 149 (resistant) has also been found economical in Uttarakhand.
- Resistance sources against finger millet blast are rare in India and Nepal, however, GPU-28 and GPU-48 is widely used cultivar highly resistant to neck and finger blast.

Biological Control

- Treat the seed with bioagents like *T. harzianum* or *Pseudomonas fluorescens* @ 6 g/kg and two sprays of *P. fluorescens* @ 0.3% first at flowering and second 10 days later.
- Spray *P. fluorescens* @ 6 g/L of water. First spray immediately after noticing the symptom. Second and Third sprays at the flowering stage at 15 days interval.
- Treat the seed with *P. fluorescens* @ 6 g/kg seed and spray the extracts of *Prosopis juliflora* leaf extract (10%), *Ipomoea carnea* leaf extract (10%).
- Foliar spray with pre-mixture fungicide (Carbendazim + Mancozeb) @ 0.2% concentration at 50% earhead emergence followed by a second spray with *P. fluorescens* at 2 g/L of water 10 days later.
- Choose a resistant cultivar coupled with seed dressing with Carbendazim (@ 2 g/kg of seed) or bioformulation of *P. fluorescens* (6 g/kg of seed).

Chemical Methods

- Seed treatment with Carbendazim @ 1 g/kg of seed.
- Spray Carbendazim 0.1% in the nursery 10–12 days after sowing. Repeat the spray 20–25 days after transplanting and 40 to 45 days after transplanting.
- Spray fungicide such as Carbendazim 500 g or Iprobenphos (IBP) @ 500 ml/ha or premixture of fungicide (Carbendazim + Mancozeb) @ 500 g/ha. Give first the spray

immediately after noticing the symptoms and second and third sprays at the flowering stage at 15 days interval.

- Treat seed with Tricyclazole (8 g/kg of seed) followed by two sprays of any one of the following combinations: (i) Ediphenphos or Kitazin or Propiconazole (0.1%) or Carbendazim or Tricyclazole (0.05%), first at the time of ear emergence and second after 10 days, (ii) first spray of Carbendazim or Tricyclazole @ 0.05%, and second spray of Mancozeb @ 0.2% 10 days later.
- Give two sprays of Carbendazim + Mancozeb (0.2%) or Carbendazim 0.05% or Tricyclazole 0.05% with first spray at 50% flowering followed by the second 10 days after.

8.2 LEAF SPOTS

8.2.1 Cercospora Leaf Spot

Cercospora leaf spot is one of the major and most destructive foliar diseases of finger millet although restricted to certain geographical regions. It occurs in the Himalayan foothills and mid-hills of Nepal. It can cause up to 40% yield losses if the disease occurs just after ear emergence.

Symptoms

Initial symptoms of the disease appear as reddish-brown specks surrounded by a yellow halo. The lesions enlarge in size and several lesions coalesce to form large lesions usually surrounded by a yellowish halo. The fungus sporulates and produces greyish-white growth in the center of these lesions, during rains or under high relative humidity conditions. These lesions enlarge and appear as eye or diamond-shaped spots and can be easily confused with the lesions of the blast. If infection occurs very severe leaves turn necrotic, dry, and shrivel and plants look completely blighted. Similar symptoms of this disease also appear on the leaf sheath, stem, peduncle (neck), and fingers (head).

Causal Organism and Etiology

Cercospora leaf spot of finger millet is caused by *Cercospora eleusinis*. The fungus produces conidiophores and conidia on the stroma. Conidiophores are hyaline, thread-like, straight to slightly curved, and 3–12 septate.

Disease Cycle and Epidemiology

The disease is generally confined to the mid-hills of Nepal, where rainfall is generally high and the average daily temperature does not exceed 20°C. However, in India, the disease is found to be a serious problem in the mid and high hills of Uttarakhand and is known to occur from 850 m to >1900 m altitude while the intensity of the disease was reported to be low in lower hills. The disease appears most severely during June in the early sown crops.

Management

- Collect and destroy the crop debris.
- Use pathogen-free seeds for sowing.

- Grow disease resistant or tolerant varieties.
- Give two foliar sprays of Carbendazim @ 0.05% at 15 days intervals.

8.2.2 Brown Spot or Leaf Blight

The brown spot disease was first reported by Butler (1918) causing seedling blight or leaf blight, foot rot of Ragi. The disease is known to be prevalent in most parts of the finger millet growing areas in India.

Symptoms

The pathogen affects almost every plant part such as the base, culms, leaf sheath, leaves, peduncle, and fingers. The characteristic symptoms on the young leave appear as minute, oval, light brown lesions on the leaves and later become dark brown. Under high humid conditions, woolly growth of the fungus may appear in the center of these lesions. Several lesions coalesce to form large patches of infection on the leaf blade. The affected blades wither prematurely and the seedlings may be killed. Linear oblong and dark brown spots appear on the leaves of older plants. The leaves give a blightening appearance. Neck infection causes dark tan lesions which initially enlarge and extend both up and downwards from the neck. In severe cases, the peduncle breaks and hangs down from the plants. The pathogen may infect earhead, fingers, and grains. Severe infection of earhead causes poor development of grain, shriveled, chaffy and discoloured seeds, resulting in heavy crop losses. If infected seeds are used for sowing, pre- or post-emergence rotting of the seedlings can be seen.

Causal Organism and Etiology

Brown spot of the finger millet is caused by *Drechslera nodulosa* (Syn. *Helminthosporium nodulosum*) [Perfect stage: *Cochliobolus nodulosus*]. The mycelium is intra- and inter-cellular septate and light brown in colour. Conidia are erect or curved septate and dark brown. Conidiophores are thick-walled cylindrical or obcavate straight or curved, light green and 3–10 septate. The spores germinate either through the stomata or epidermal cells.

Disease Cycle and Epidemiology

Brown spot disease of finger millet is seed-borne and primary infection occurs through seed. The conidia carried on the seeds can remain viable for one year. The fungus also remains viable on crop stubbles and the dead host remains and is able to survive in the soil for 18 months. The first infection is caused on the seed by the fungus. Then the fungus becomes systemic from the early stages. Secondary infection occurs through airborne conidia. Infection can occur between temperatures 10°C–37°C, however, the optimum is 30°C–32°C. Nursery infection can also cause heavy losses due to the seedling blight, and neck infection. Several other hosts plants are also infected by this pathogen including *Echinochloa frumentacea, Eleusine indica*, *Setaria italica*, *Panicum miliaceum, Dactyloctenium aegyptium*, *Pennisetum typhoides*, *Zea mays*, and *Sorghum vulgare.*

Management

- Use pathogen-free seeds for sowing.
- Remove crop residues, weed hosts, and others host from the fields.
- Grow disease resistant or tolerant varieties.
- The disease is primarily seedborne therefore treat the seed with Captan or Thiram @ 4 g/kg of seed.
- Spray Mancozeb @ 0.2% if needed.
- Spray Bordeaux mixture @ 1% or Copper oxychloride or Dithane Z-78 @ 2 g/L of water.

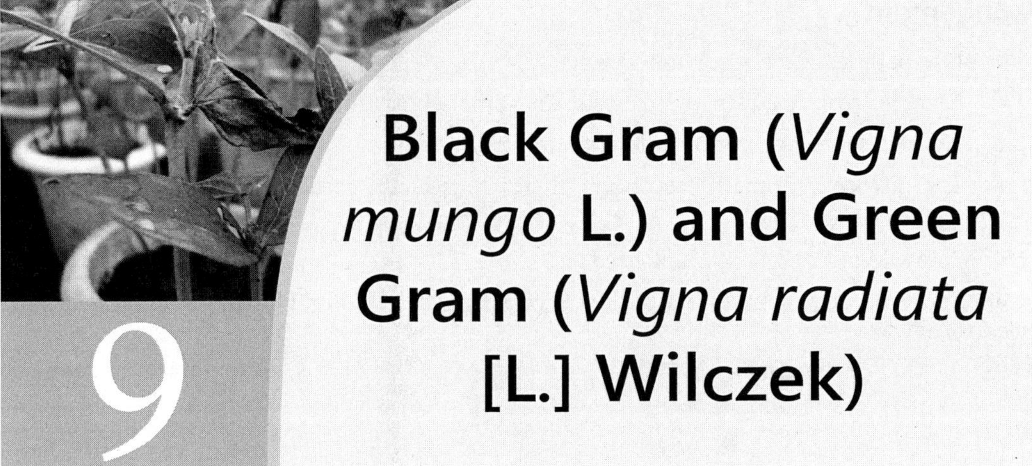

9 Black Gram (*Vigna mungo* L.) and Green Gram (*Vigna radiata* [L.] Wilczek)

9.1 CERCOSPORA LEAF SPOT

The Cercospora leaf spot was for the first time reported from Delhi, India. Now the disease is prevalent worldwide and causes huge yield losses by causing severe leaf spotting and defoliation. Yield losses due to this disease have been observed up to 50%–70%. However, under field trials, a wide range of yield losses (23%–96%) due to this has been reported from different states of India.

Symptoms

The characteristic symptom appears as numerous small, water-soaked, round or angular spots with pale brown center and reddish purple margin on leaves, branches, petioles, and pods (Figure 9.1). During flowering and pod formation severe leaf spotting and defoliation occurs. On the pods, symptoms range from a few spots to solid blacking and killing of the entire pod and shriveled, small, and discoloured seeds within the infected pod.

Figure 9.1 Cercospora Leaf Spot of Mungbean.

Causal Organism and Etiology

Cercospora leaf spot of mungbean and urdbean is caused by *Cercospora* spp. viz., *C. cruenta* (Perfect state: *Mycospharrella cruenta*), *C. canescens*, *C. kikuchii*, *C. dolichi* and *C. corocollae*. Two species *C. cruenta*, and *C. canescens* are frequently observed on mungbean. The fungus produces clusters of dark brown septate conidiophores and conidia are linear, hyaline, thin walled and 5–6 septate.

Disease Cycle and Epidemiology

The fungus survives on diseased plant debris and seeds. Seedlings from infected seeds produce cotyledonary lesions. Sporulation on the cotyledonary lesions provides primary inoculum for infection of young leaves. Under humid weather condition, sporulation is profuse. Secondary infection is initiated by conidia under warm and wet conditions which are disseminated by wind and rain splash to other leaves stems and pods. The disease development is favoured by a temperature range 25°C–30°C with high relative humidity (90%–100%). Moist weather and splattering rains are conducive to disease development. Most outbreaks of the disease can be traced back to heavy rainstorms that occur in the area.

Management

Cultural Methods

- Use pathogen-free seeds.
- Field sanitation.
- Intercrop the mungbean or urdbean with tall growing cereals and millets.
- Crop rotation with nonhost crop.
- Remove and destroy infected crop debris of the previous crops.
- Avoiding planting of collateral hosts in the vicinity of the crop.
- Mulching reduces the disease incidence resulting in increased yield.
- Maintain low crop population density and wide row planting.

Host Resistance

- Grow resistant/tolerant varieties. (Mungbean: LM 113, LM 168, LM 170, JM 171; Urdbean: Naveen, Jawahar Urd-3, Gujarat Urd-1, and Barkha)

Biological Control

- The crude extracts of cassava, garlic, and zinger are applied for controlling the disease effectively.

Chemical Control

- Treat the seeds with Thiram or Captan @ 2.5 g/kg of seed.
- Two foliar sprays of Difenoconazole (0.0125%) or Carbendazim (12%) + Mancozeb (63%) – 75 WP (0.0.2%) or Hexaconazole (5%) + Captan (70%) – 75 WP (0.05%) or Ridomil Gold (0.1%) or Propiconazole (0.1%) are effective for minimising disease incidence, severity, and increasing yield.

- Spray Mancozeb or Copper oxychloride @ 2 kg/ha or Carbendazim or Benomyl @ 500 g/ha.

9.2 ANTHRACNOSE

The disease causes both qualitative and quantitative losses. A wide range of yield losses (24%–67%) due to anthracnose disease has been estimated from several mungbean growing areas in India.

Symptoms

The anthracnose may appear on all foliar plant parts including leaves, petioles, stem, and pods. Initially, water-soaked lesions appear on leaves and pods which later become circular, black, sunken spots with a dark center and bright red-orange margins on leaves and pods. In severe cases, the affected parts wither off.

Causal Organism and Etiology

Anthracnose of mungbean and urdbean is caused by *Colletotrichum* spp. viz. *Colletotrichum lindemuthianum* (perfect state: *Glomerella lindemuthianum) C. dematium, C. capsici, C. trucatum*, and *C. graminicola*. All these five species of *Colletotrichum* are known to attack mungbean and urdbean. The mycelium of the pathogen is septate, hyaline, and branched. Conidia are produced in acervuli. The conidiophores are hyaline and short and bear oblong or cylindrical, hyaline, thin-walled, single celled conidia with oil globules. The sexual stage of the fungus produces perithecia with the limited number of asci, which contain typically 8 ascospores either one or two-celled with a central oil globule.

Disease Cycle and Epidemiology

The fungus is seed-borne and causes primary infection. It also perpetuates in the infected plant debris in the soil. The secondary spread by airborne conidia produced on infected plant parts. The pathogen is disseminated by the spattering rains splash associated with wind. The pathogen infects several other leguminous hosts that may also serve as reservoirs of the primary inoculum. The disease is more severe in cool and wet seasons. The disease development is favoured by high humidity (100%) and moderate temperature (17°C–26°C), and intermittent rains at frequent intervals.

Management

Cultural Methods

- Use disease free seed.
- Follow crop rotation.
- Remove and destroy the infected plant debris from the field after harvesting the crop.

Physical Control

- Hot water treatment of seed at 58°C for 15 minutes is effective for controlling seedborne infection and increasing seed germination.

Host Resistance

- Grow resistant or tolerant varieties, e.g. mungbean genotypes/cultivars viz., TM-92-2, TARM-18, MLTG-9.

Chemical Control

- Treat the seeds with Thiram 80% WP @ 2 g/kg or Captan 75 WP @ 2.5 g/kg Carbendazim @ 2 g/kg.
- Treat the seeds with the combination of Thiram + Carbendazim (2:1) @ 3 g/kg of seed.
- Spray Carbendazim 500 g or Mancozeb 2 kg/ha soon after the appearance of the disease and repeat after 15 days.

9.3 WEB BLIGHT

It is one of the major constraints in many pulses production in warm humid tropic zones of the world. On mungbean, Rhizoctonia blight was reported for the first time from Philippines (Nacien, 1924) in 1924. In India, Dwivedi and Saksena (1974) first reported its occurrence on mungbean from Kanpur (Uttar Pradesh). In Pakistan, Alam et al. (1985) reported the occurrence of web blight of mungbean. In Rajasthan, yield loss due to web blight has been reported about 30%–40% and from Jabalpur, central India, 20%–40% seedling mortality due to *Rhizoctonia* infection has been reported.

Symptoms

The symptoms of web blight of mungbean and urdbean can be observed as leaf and pod spots, leaf blights, defoliation, stem and petiole lesions, cob-web like mycelium and produces abundant microsclerotia and macrosclerotia on diseased plant tissue during the growing season (Figure 9.2).

Figure 9.2 Web Blight of Mungbean. (A) Mycelial web of *Rhizoctonia solani* on leaves, (B) Severe blight symptoms with mycelial web and sclerotia on infected plant parts.

Causal Organism and Etiology

Web blight of mungbean and urdbean is caused by *Rhizoctonia solani* Kühn (Teleomorph: *Thanatephorus cucumeris*). The rapidly growing mycelium of *Rhizoctonia solani* are hyaline when young but later turn brown; branching near the distal septum of the hyphal cell, often nearly at right angles in the older hyphae. Construction of branched hyphae at the point of origin and possess septum near its junction with the main axis. The cells of growing hyphae are always multinucleate and the septum is dolipore type. Sclerotia are white but later turn chestnut brown, globose to subglobose, oval or cushion shaped, often less than 1 mm in diameter to thin crusts several centimetres across.

Disease Cycle and Epidemiology

The epidemiology of Rhizoctonia web blight may be divided into two phases, one before and the after canopy closures. The first phase of epidemiology is soil-borne and the second is leaf-borne. The pathogen is primarily soilborne and can survive for many years by producing sclerotia in soil and on infected plant tissues/seed coats. *R. solani* also survive as saprophytes by colonizing soil organic matter. The secondary spread of web blight disease is by mycelial bridges between plants, rain splashed sclerotia, infested soil debris and airborne basidiospores. The primary inoculum of *R. solani* germinates to produce vegetative threads (hyphae) of the fungus that can attack mungbean onto the leaves during heavy rains. The fungus is attracted to the plant by chemical stimulants released by actively growing plant cells and/or decomposing plant residues. As the fungus kills the plant cells, the hyphae continue to grow and colonize dead tissue, often forming sclerotia. The pathogen also infects floral parts and seeds. The collateral weed hosts play an important role in the initiation and early spread of the disease to the main host. Weed hosts first infected and facilitates the production of the basidiospores of the pathogen. Later the pathogen become airborne and cause leaf, stem, or pod infection on the main host and produces fresh crops of basidiospores. Maximum production and discharge of basidiospores occur during midnight and early morning before sunrise. Night temperature below 24°C and relative humidity above 95% favour spore production. In North India, basidiospore production starts in mid August under natural conditions. The basidiospores germinate in 2 hours and penetrate the leaf through the intact surface by the formation of an infection cushion or direct penetration through stomatal openings. Severe web blight disease development is favoured by higher aerial temperature (26°C–32°C), high relative humidity (100%), and soil temperature 30°C–33°C. During the early stage of the crop in urdbean, rainfall (90–100 mm) also plays a significant role in the severe development of web blight.

Management

Cultural Methods

- Use pathogen-free seeds for planting.
- Early sowing of plants (upto July 1st).
- Avoid dense planting.
- Completely cover plant residue by clean ploughing the field soon after harvest.
- Destroy infected stubble.
- Deep summer plowing.

- Improved drainage help in reducing the disease.
- Burying the infected leaves immediately after the harvest.
- Remove weed hosts.
- Shallow planting also helps in minimising the infection by *R. solani.*
- Wide row spacing or lower plant populations help in the reduction of aerial blight.

Host Resistance

- Use resistant/tolerant varieties.

Chemical Control

- Treat the seeds with Thiram + Carbendazim (2:1) @ 2.5–3 g/kg seed. Seed treatment limits early season disease development.
- Spray Carbendazim @ 1.0 g/L of water, 45–60 days after sowing.

9.4 YELLOW MOSAIC DISEASE

In India, Mungbean yellow mosaic virus (MYMV) was first reported by Nariani in 1960 from the mungbean fields of IARI, Pusa, New Delhi. In mungbean and urdbean, yellow mosaic disease is of key importance, especially in South and Southeast Asia. The overall crop yield loss may range between 10%–100%, depending on the mungbean genotype and stage of crop infection. Yellow mosaic intensity of 25% and above influences pod formation and yield in urdbean. This virus disease is the most destructive disease of Kharif (summer) legumes in India.

Symptoms

Symptoms of the disease are almost similar in mungbean and urdbean and other host plants also. Severe yellow mosaic and mottling of leaves, malformation, size reduction, and stunting of plants are the major symptoms. Symptoms first appear as small yellow flecks or specks or spots and irregular blotching in young leaves (Figure 9.3). Subsequently emerging leaves

Figure 9.3 Yellow Mosaic of Mungbean.

exhibit more conspicuous and irregular yellow and green patches alternating with each other. In urdbean, two types of symptoms viz., yellow mottle and necrotic mottle are considered to be caused either by different strains of the same virus or by different viruses. Yellow spots may also be observed on pods and green seed coats of developing seeds. The pods are deformed and contain shriveled and undersized seeds. The number and size of pods per plants and seed per pod are reduced greatly.

Causal Organism and Etiology

Yellow mosaic disease (YMD) of mungbean and urdbean is caused by *Mungbean yellow mosaic India virus* (MYMIV) in the Northen and Central region and *Mungbean yellow mosaic virus* (MYMV) in the western and southern regions in India. However, at present three distinct viruses viz., *Mungbean yellow mosaic India virus* (MYMIV), *Mungbean yellow mosaic virus* (MYMV), and *Horsegram yellow mosaic virus* (HgYMV) are known to cause YMD in mungbean and urdbean in India. These are Begomovirus belonging to the family geminiviridae. Germinate virus particles, ssDNA, bipartite genome with two genomic components DNA-A and DNA-B.

Host Range: Besides mungbean and urdbean, YMD also affects various leguminous crops including mothbean (*Vigna aconitifolia*), Lima bean (*P. lunatus*), pigeon pea (*Cajanus cajan*), French bean (*Phaseolus vulgaris*), cowpea (*Vigna unguiculata*), Dolichos (*Lablab purpureus*), horsegram (*Macrotyloma uniflorum*), and soybean (*Glycine max*).

Disease Cycle and Epidemiology

A number of cultivated legume plant species and weeds (*Croton sparsiflorus, Acalypha indica, Eclipta alba*) serves as a reservoir for inoculum, and inoculum is available round the year of yellow mosaic disease causing viruses. However, the prevalence and severity of the disease vary and depend on several factors such as the prevalence of alternate or collateral hosts of the virus(es) nearby the crop and the availability and population of vector (whitefly). The virus(es) is/are transmitted by whitefly, *Bemisia tabaci*, nonpersistently. The virus is not transmitted to the offspring of the viruliferous whitefly indicating no transovarial transmission. These are not transmitted through sap and pollen or seed. Whiteflies may acquire the virus within 15 minutes of the acquisition feeding period, and it requires an incubation period (virus in vector) of 4 hours before being able to transmit the virus. A minimum of 15 minutes (inoculation feeding period) is required to transmit the virus to a healthy host plant. The virus enters the phloem cells of the host through the whitefly proboscis and the viral aggregates appear in the host cell nuclei roughly two days before the symptom appearance. A single whitefly can transmit the virus. Female whitefly is more efficient than male in transmitting the virus. Female whitefly may retain the virus for upto 10 days while male whitefly generally retains the virus for upto 1 day.

Management

Cultural Methods

- Rogue out the diseased plants up to 40 days after sowing to prevent the further spread of the disease.
- Remove the weed hosts periodically.

- Increase the seed rate (25 kg/ha).
- Cultivate the crop during thc *Rabi* season.
- Plant borders of crops such as sorghum, pearl millet or maize around mungbean/urdbean crops.
- Follow intercropping/mixed cropping. Grow two rows of maize or sorghum or cumbu for every 15 rows of black gram or green gram.

Host Resistance

- Grow resistant green gram varieties like Pant Mung, Pusa 105, Pusa Vishal, Basanti, ML-5, ML-337, PDM-54, Samrat, PDM 54 (Moti), PDM 84-139 (Samrat), PDM 84-143, PDM–11, ML–337, MUM 2, MH-88-111, MUM-2, Narendra Mung 1, Pant Mung 3, PDM 139 (Samrat) (Spring season), ML 131, and ML 267.
- Grow resistant/tolerant varieties of black gram like IPU 94-1 (Uttara), Shekhar 3, KU 309, Ujala (OBJ 17), VBN (Bg)7, Pratap Urd 1, Pant U 19, Pant U–30, UG 218, PDU 1, PDU 88-31, Narendra Urd 1PS1, WBU 108, KU 92-1 (Spring season), KU 300 (Spring season), etc.

Biological Control

- Use admixture of neem seed kernel extract (in cow urine) @ 3% + Dimethoate @ 0.03% and of neem oil @ 0.5% + Dimethoate @ 0.03% to be effective in reducing the incidence not only of whitefly and yellow mosaic but also of pod borer.
- Spray different formulations of neem oil @ 3%, and neem seed kernel extract @ 5% at 35 and 55 days after sowing.

Chemical Control

- Treat the seeds with Thiomethoxam 70 WS or Imidacloprid 70 WS @ 4 g/kg.
- Spray Thiamethoxam 25 WG @ 100 g or Imidacloprid 17.8% SL @ 100 ml in 500 L of water.
- If required spray with Triazophos 40 EC @ 2.0 ml/L or Oxydemeton methyl 25 EC @ 2.0 ml/L or Malathion 50 EC @ 2.0 ml/L of water at 10–15 days intervals to prevent whitefly infestation.

10 Castor (*Ricinus communis* L.)

10.1 PHYTOPHTHORA BLIGHT

Phytophthora seedling blight is a serious disease of germinating seed, affecting cotyledonary leaves and growing points, causing plant mortality resulting in loss of plant stand. In 1913, J.F. Dustur, discovered Phytophthora blight disease of the castor oil plant and establish a new species of *Phytophthora* and named *P. parasitica.*

Symptoms

The disease appears circular, dull green patch on both the surface of the cotyledonary leaves. Later on, the spots turn yellow or brown with concentric lighter and darker brown zones on the lower surface of the leaves and stems. The infection spreads to the stem and causes withering and death of seedlings. In mature plants, the infection initially appears on the young leaves and spreads to petioles and stem causing black discolouration and severe defoliation. Under humid condition, the growth of the pathogen appears as a white web on under the surface of the leaves. The most severe damage occurs with the infection of the growing point of leaves with rotting.

Causal Organism and Etiology

Phytophthora blight of caster is caused by *Phytophthora parasitica*. The pathogen produces non-septate and hyaline mycelium. Sporangiophores emerge through the stomata on the lower surface singly or in groups. They are unbranched and bear single-celled, hyaline, round or oval sporangia at the tip singly. The sporangia germinate to produce abundant zoospores. The fungus also produces oospores and chlamydospores in adverse seasons.

Disease Cycle and Epidemiology

The pathogen remains in the soil as chlamydospores and oospores which act as the primary sources of infection. The fungus also survives on other hosts like potatoes, tomatoes, brinjal, sesamum, etc. The secondary spread takes place through wind-borne sporangia. Favorable environmental conditions that favour the disease development are continuous rainy weather, low temperature (20°C–25°C), low-lying, and poorly drained soils.

Management

- Remove and destroy infected plant residues.
- Avoid low-lying and poorly drained fields for sowing.
- Treat the seeds with Thiram or Captan @ 4 g/kg.
- Do seed dressing first with Metalaxyl @ 3 g/kg and then with bioformulation of *Trichoderma viride* @ 10 g/kg of seed.
- Soil drenching with Copper oxychloride @ 3 g/L of Metalaxyl @ 2 g/L of water.

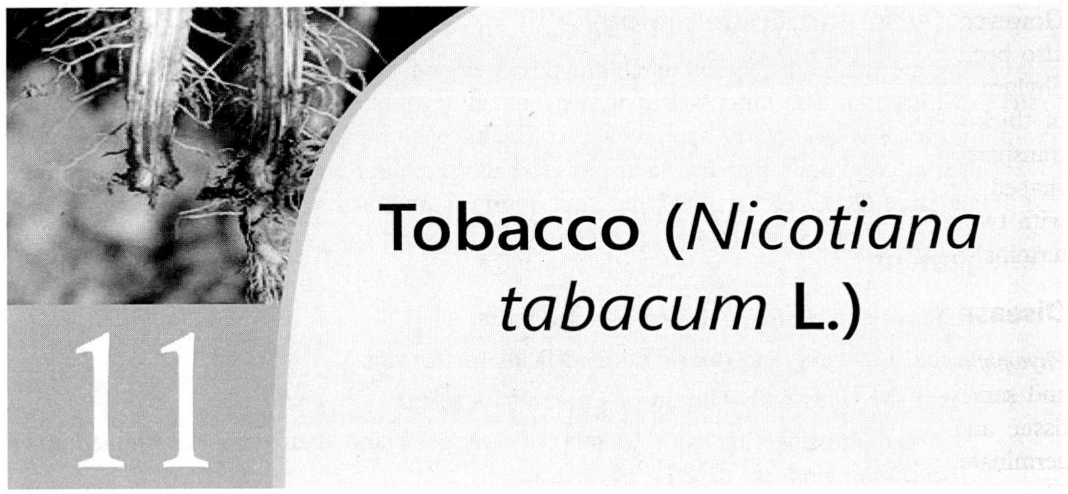

11 Tobacco (*Nicotiana tabacum* L.)

11.1 BLACK SHANK

Black shank is a devastating disease of all types of tobacco. It causes losses in individual wet fields reaching upto 100%. The disease was first described from Indonesia (1896) and is now reported from most major tobacco-growing areas worldwide.

Symptoms

Plants of all ages may be infected, but seedlings are most susceptible.

Above-ground symptoms: Above-ground symptoms of the stem and root rot phases of the disease begin as a temporary wilt that progresses into chlorosis of leaves and permanent wilting. Sometimes, the pathogen can attack the leaves, producing large water-soaked spots, which enlarge to blight the leaves. One of the most characteristic symptoms of the black shank is seen when the stem is split or cut longitudinally. The pith is dry, brown to black, and separated into plate-like disks (necrotic disking of the pith). In addition, the infection also can occur directly on the stem which results in a stem lesion with little or no root necrosis. Such symptoms are sometimes referred to as "stem black shank" and may result in plant death.

Below ground symptoms: The most common symptom of the disease is the root and crown rot. Infected root tissue initially develops water-soaked, then rapidly become necrotic, and appears dark brown to black elongated lesions. In susceptible varieties, colonization of root tissues spread rapidly and reaches upto the stem, resulting in the characteristic black shank symptom and plant death.

Causal Organism and Etiology

Currently, *Phytophthora nicotianae* is the most frequently used name for the black shank pathogen.

However, *Phytophthora parasitica* var. *nicotianae* and *Phytophthora nicotianae* var. *nicotianae* have also been used. It is a fungus-like organism in the Kingdom Straminipila (or Stramenopila), phylum Oomycota; heterothallic, requiring two mating types (A1 and A2) for the production of thick-walled oospores; antheridium is amphigynous; diploid; hyphae hyaline (colourless, transparent), aseptate or coenocytic, few to numerous hyphal swellings; sporangia ovoid, pear-shaped, or spherical, and have very conspicuous papillae; zoospore typically kidney-shaped, with two flagella (posterior whip-like and anterior tinsel type); chlamydospores thick-walled, terminal or intercalary.

Disease Cycle and Epidemiology

Phytophthora nicotianae survives as mycelia and chlamydospores in infected crop residue in soil and serves as the primary inoculum that initiates infection. Oospores overwinter in infected tissue and soil, but their function is not known. In warm moist soil, the chlamydospores germinate and produce germ tubes that either directly infects the plant or produce a sporangium. The sporangia then germinate and form either hyphae or zoospores. Zoospores and sporangia germinate and produce hyphae that infect plants. Zoospores encyst, germinate, and penetrate roots within one hour. Sporangia serve as secondary inoculum and can form within 24 hours. Sporangia are disseminated by wind and water that initiate further infections on host plants. Black shank is a polycyclic disease. Root and stem colonization of *Phytophthora nicotianae* results in typical root rot and black shank symptoms. Black shank is most severe in wet soils at temperatures above 21°C. Root damage caused by plant parasitic nematodes such as *Meloidogyne* and *Globodera* can increase disease infection and severity.

Management

Cultural Methods

- Select healthy, disease-free seedlings for transplanting.
- Remove and destroy the affected plants in the field.
- Deep ploughing in summer to reduce the inoculum.
- Preparation of raised seed beds 15 cm high with channels around to provide drainage.
- Cover the seed beds before sowing with paddy husk, groundnut shell tobacco stubbles, waste grass, etc. at 15–20 cm thick layer and burn.
- To avoid over-crowding of seedlings use seed rate @ 3.5 kg/ha.
- Watering should be regulated to avoid excessive dampness on the bed surface.
- Use good management practices to manage nematodes.
- Avoid cultivation in areas or a field infested with *P. nicotianae*.
- Disinfect machinery parts with fungicides after use.
- Clean shoes after working in infested soil.

Host Resistance

- Grow resistant or tolerant varieties like Mc.Nair-12 or CM-12(KA) or K-326 which are black shank tolerant varieties are to be used in irrigated northern light soils.

Chemical Control

- Provide adequate drainage in the nursery. Drench the nursery beds with 1% Bordeaux mixture or 0.2% Copper oxychloride, two days before sowing.
- Spray Ridomil MZ 72 WP @ 20 g or Fenamodone 10% + Mancozeb 50% @ 30 g in 10 L of water twice at 20 and 30 days after germination. Ridomil should not be sprayed more than two times or more than the recommended dose and 500 g of Ridomil is sufficient for one acre of the nursery.
- Spray the beds two weeks after sowing with Metalaxyl or Captafol or Copper oxychloride @ 0.2% or Bordeaux mixture @ 1% and repeat after 10 days.
- Spray Mancozeb 2 kg or Copper oxychloride 1.0 kg or Ziram 1.0 L/ha.
- Spot drench with Copper fungicides such as Bordeaux mixture @ 0.4% or Copper oxychloride @ 0.2%.

11.2 BLACK ROOT ROT

Black root rot disease is also called maricume radicale, root rot, and Thielavia root rot. The disease is most common and most damaging in tobacco if the varieties are not resistant. The disease occurs both in traditional nurseries and in floating seedbeds.

Symptoms

Above-ground symptoms of the disease are typical of many other root rots or root problems and include chlorosis, defoliation, stunting, drooping of the leaves and wilting and discolouration of leaves and shoots, red-brown lenticels and swelling of the crown. Infected root tissue first develops dark brown to black elongated lesions. Damping off symptoms of seedlings is seen in plant beds. The characteristic root symptoms are numerous dark blackened root tips with reduced and stubby roots. Infected root tissue first develops dark-brown to black elongated lesions and quickly progress until the entire length of the root takes on a black discolouration. The cortical tissue collapses and the epidermis and cortical tissue may slough off. Young roots are also black with discrete lesions. The fungus produces black spores in roots, when abundant, causes the black lesions that are typical of black root rot.

Causal Organism and Etiology

Thielaviopsis basicola (also called *Chalara elegans*) is a soilborne fungus in the phylum Ascomycota. *T. basicola* reproduces asexually by producing two types of conidia, endoconidia and aleuriospores. Endoconidia (also called phialospores) are single-celled hyaline spores with slightly rounded ends, produced within elongate terminal phialides; the conidia may stick together end-to-end, forming long unbranched chains; aleuriospores are darkly pigmented cylindrical spores contain 2–8 terminal cells. At maturity and before germinating, the cells fragment along the transverse septa, leaving 2–8 short-cylindrical, one-celled spores, usually referred to as chlamydospores.

Disease Cycle and Epidemiology

Chlamydospores and endoconidia of *T. basicola* can survive in soil for several years and are considered to be the primary infective propagule. Chlamydospores and endoconidia germinate and form hyphae that penetrate roots. *T. basicola* is a hemibiotrophic plant pathogen, with an initial biotrophic phase as it invades and colonizes living cells followed by a necrotrophic phase with the death of the plant cells. The fungus is spread when healthy roots come in contact with infected roots or infested soil and when aleuriospores are dispersed by splashing water.

Chlamydospores and endoconidia produced in mycelium on the outside of diseased roots serve as secondary inoculum. The movement of the pathogen from field to field is attributed to the movement of infested soil on equipment. Black root rot is most severe at relatively cool soil temperatures (13°C–23°C) and wet soils. Black root rot is generally more severe at soil pH values above 5.6 and suppressed under more acidic conditions (pH below 5.2)

Management

Cultural Methods

- Change the site of the bed each year.
- Rotate tobacco with resistant crops, such as grass, for 3 years or longer.
- Avoid using decomposing residues of some green manure cover crops (i.e. barley and rye) because they may predispose roots of resistant tobacco cultivars to *T. basicola.*
- Avoid using the leguminous plants as cover crops which can increase populations of *T. basicola.*
- Treat soil with steam or chemicals if an old bed site is used.
- Avoid excessive use of fertilizer.
- Avoid transplanting in cool soil or air temperature is low.
- Maintain a soil pH of 6.1–6.5 for burley and 5.6–6.0 for dark tobacco.
- Remove and destroy all diseased plants.
- Use soil-less media if possible.
- Fungus gnats and shore flies may be vectors of the pathogen.

Host Resistance

- Grow resistant cultivars.

Chemical Control

- Soil drenching with Thiophanate-methyl or Fludioxonil or Triflumizole or Thiophanate-methyl or Etridiazole should be done as a preventive measure or at the first appearance of the disease.
- Sterol-inhibiting fungicides like Flusilazole and Triadimenol also have been shown to be effective in disease management in field crops.

11.3 TOBACCO MOSAIC

It was the German Adolf Mayer, working in the Netherlands in 1882 first described an important disease of tobacco which he called tobacco mosaic disease. Stanley, in 1935, at the Rockefeller Institute, purified *Tobacco mosaic virus. Tobacco mosaic virus* (TMV) was the first virus discovered. In 1889, Martinus Beijerinck, found that 'tobacco mosaic disease' was caused by a pathogen able to reproduce and multiply in the host cells of the plant and called it *virus* (meaning poison). It was the first virus ever purified. In 1935, Stanley purified TMV and derived fine needle-like crystalline preparations which were fully active and infectious. Tobacco yield losses due to TMV are currently estimated at only 1%, because resistant tobacco varieties are routinely grown. However, TMV affects other crops, and losses of up to 20% have been reported in tomatoes.

Symptoms

Specific symptoms depend on the host and age of the plant, virus strain and genetic background of the host plant and environmental conditions. However, some common symptoms include mosaic-like patches (light and dark green patches), mottling and necrosis on the leaves, leaf curling, yellowing of plant tissues, and stunting of plants. Dark-green blisters and sometimes enations (leafy growth) appear on the dorsal side of the leaf. In case of severe infections, the leaves are narrowed, puckered, thin, distorted, and malformed which gives "shoestring" effect. Under hot weather, dark brown necrotic spots develop, and this symptom is called "Mosaic burn" or "Mosaic scorching".

Casual Organism and Etiology

Tobacco mosaic virus (TMV) is within tobamovirus group. The rod-shaped virus particles (virions) of TMV measure about 300 nm × 15 nm. A single TMV particle is composed of 2,130 copies of the coat protein (CP) that envelope the RNA molecule of about 6,400 nucleotides. The coat protein of the virus contains 158 amino acid residues and was the first viral protein to be sequenced and to have its three-dimensional structure elucidated by X-ray crystallography.

Hosts Range: Tobacco, tomato, and other solanaceous plants.

Disease Cycle and Epidemiology

TMV overwinters in several perennial weeds (horse nettle *Solanum carolinense*, ground cherry, *Physalis angulata*), air-dried tobacco, and infested residue in the field. Air-dried tobacco is a common source of new infections. TMV is disseminated mechanically and is not known to transmit by an insect vector. The virus is also not seed-transmitted in tobacco but tomato seeds transmit the virus. TMV is highly contagious and transmitted most rapidly and very easily by contact wounds (rubbing of infected leaf against a leaf of a healthy plant), sap and contaminated farm implements, clothing, and occasionally by workers whose hands are contaminated with TMV after smoking cigarettes. Agricultural practices, such as continuous cropping can enhance TMV inoculum in many host plants. The virus has a wide host range, affecting nearly 50 plant species belonging to nine different families. The purified virions of TMV are highly stable and remain infectious after 50 years of storage at 4°C in the laboratory. TMV particles are released from infected plants, when plant cells are ruptured or the dried leaves are blown as "dust". These particles infect new plants through wounds. Initially, the virions reach the vascular system (veins) and later become systemic through the phloem.

Management

Cultural Methods

- Use virus-free plants.
- Propagate plants via seed rather than vegetative propagation.
- Wash hands thoroughly after handling tobacco products or TMV-infected plants.
- Remove all crop debris from fields, benches, and greenhouse.
- Workers should disinfect their hands with soap and running water before weeding or handling seedlings.
- Wash tools with soap or 10% solution of household bleach to inactivate the virus.
- Workers should not chew or smoke while working in the nursery.
- Spray the plant bed with milk 24 hours before uprooting seedlings. Workers should dip their contaminated hands in milk to inactivate TMV in every 20 min. while working with seedlings and prior to planting.
- Rotate tobacco in a field every 2 years.
- Rogue out diseased plants and destroy them.
- Weeds (*Solanum nigrum*) and plants (brinjal, tomato, chillies) susceptible to the virus should be destroyed.
- Phytosanitary measures are very important for managing this disease.

Host Resistance

- Grow resistant cultivars such as Rajahmundry, viz., TMVRR-2, TMVRR-2a, and TMVRR-3 to overcome this problem.

Biological Control

- Spray leaf extracts of Bougainvillea (*Basella alba*) @ one litre of extract dissolved in 100–150 litres of water on 30th, 40th and 50th days after transplanting.

12 Guava (*Psidium guajava* L.)

12.1 WILT

Guava wilt disease was first reported in Taiwan by Kurosawa (1926). In India, Das Gupta and Rai (1947), first time recorded wilt disease from the guava orchards of Lucknow (UP) in severe form. Singh and Lal (1953) estimated that 5%–15% of the trees died due to wilt every year in 12 districts of UP. In West Bengal, the disease reduced the yield by 80%.

Symptoms

The affected plants show light yellow foliage with loss of turgidity and epinasty, slight leaf curling at the terminal branches, later become reddish and subsequently shedding of leaves, bare twigs, and fail to produce new leaves or flowers and eventually dry up. Affected branches produce underdeveloped, hard, black, and stony fruits. The entire plant becomes defoliated and eventually dies. Later, wilted plants show bark splitting. Finer roots show prominent black streaks, rotting at the basal region, and bark easily detachable from the cortex. The cortical regions of the stem and root show damage and distinct discolouration. Light brown discolouration of vascular tissues. Young and old fruit-bearing trees are attacked by the pathogen but older trees are more prone. Some plants may show wilting of variable degrees (yellowing and drooping of leaves, drying of terminal branches or partial wilting) during different months but later escape or resist wilting. Few plants also show partial wilting, which is a very common symptom of wilt in guava. In partial wilting, during the first year, one side of the few branches wilt and in the next year full plant dies. The rate of symptom development varies and generally wilt affected plants show two types of symptoms, i.e. slow wilt, and sudden wilt.

Slow wilt: In slow wilt plants take a year or more for complete wilting. When the disease progression is slow, usually the entire tree is affected, sectors are uncommon, shoot growth stops, subterminal leaves distorted, yellowing and defoliation occur.

Sudden wilt: Sudden wilt can be termed as true wilt and plants take 2–4 weeks to complete wilting. In complete wilting, all leaves wilt and dry, imparting a scorched appearance to the guava tree. Development of fruit is arrested and they mummify on the tree. Blisters develop in the bark of dead wood which contains masses of white to salmon-pink spores. Fast wilting can also occur in sectors, giving rise to trees with dead and healthy branches. All the affected trees die quickly.

Causal Organism and Etiology

Fusarium oxysporum f. sp. *psidii* and *F. solani* are generally the main cause of guava wilt. However, other fungi are also reported to cause wilting such as *Macrophomina phaeseoli*, *Rhizoctonia bataticola*, *Cephalosporium* sp., and *Gliocladium roseum*.

Disease Cycle and Epidemiology

The disease is soilborne. The fungus survives in infected roots of dead trees through chlamydospores. The infection occurs through root hairs or wounds or natural openings or openings caused by secondary roots. The disease is more common in alkaline soils with moderate soil moisture. The disease severity is high at pH ranging from 7.5 to 9.0, while high disease incidence in lateritic soils is at pH 6.5. Long distance spread of the disease is through the movement of plants containing sick soil in non-infested areas while short-distance spread is by water. Root injury predisposes wilt disease. Favorable conditions for disease development are maximum and minimum temperature ranges of 23°C–32°C with 76% relative humidity, high rainfall during August or September, and stagnation of water in the guava field for a long duration.

Management

Cultural Methods

- The disease can be kept under check by proper sanitation in the orchard.
- Wilted trees should be uprooted and burnt, and a trench should be dug around the tree trunk. While transplanting, the roots of plants should not be damaged severely.
- Maintenance of proper tree vigour in timely and adequately.
- Manuring, intercutlural and irrigation enable them to withstand infection.
- The pits may be treated with formalin, and kept covered for about 3 days and
- Transplanting should be done after two weeks.
- Organic manures, oil cakes, and lime also check the disease.
- Partially wilted plants recovered fully after rejuvenation/heavy pruning of the tree.

Host Resistance

- Use of rootstocks resistant to wilt could be an alternative effective method for the control of disease. Cross of *Psidium malle* × *P. guajava* has been found free from wilt and this material can be used as resistant root stock.

Biological Control

- An ecofriendly approach to manage the guava wilt is suggested where biological control, soil amendment, and intercropping are effective. Biological control by *Aspergillus niger* strain AN-17 is found effective.
- *Aspergillus niger* multiplied in FYM @ 5 kg/pit, applied in the pits while planting new plants. In older plants, *Aspergillus niger* enriched FYM can be applied @ 10 kg plant.

12.2 ANTHRACNOSE

Anthracnose in guava is reported to be a serious disease. It is the most common pre- and post-harvest fruit disease in all guava-growing countries. This disease is a serious problem in Uttar Pradesh, Punjab, and Karnataka. It was first reported by Mehta (1951) from Uttar Pradesh. It causes dieback, twig blight, wither tip or fruit spots. It can cause considerable postharvest losses. This disease can cause considerable postharvest losses and can affect young developing flowers and fruit.

Symptoms

Dieback phase: The plant begins to die backward from the top of a branch. Young shoots, leaves, and fruits are readily attached, while they are still tender. The fungus develops from the infected twigs and then petiole and young leaves. These may droop down or fall leaving the dried twigs without leaves. In moist weather, numerous black dots (acervuli) are produced on the dead parts of the twigs.

Fruit and leaf infection phase: The initial symptoms of anthracnose on fruit appear as minute, pin-head spots on unripe fruits which gradually enlarge. The fruits are usually infected at early maturity, but the fungus remains quiescent until symptoms develop during ripening. Spots are dark brown, sunken, circular, and have minute black stromata (acervuli) in the center of the lesion, which produce creamy or pinkish masses of spores under moist and humid weather conditions (Figure 12.1). Later, these lesions often coalesce to form larger patches and extend into the flesh. On unripe fruits, the infected area becomes corky and hardy and often develops cracks in case of severe infection. Acervuli are formed on fruit stalks. The disease may cause the shedding of unopened buds and flowers. On leaves, the fungus causes necrotic lesions at the tip or on the margin. Usually, these lesions are ashy grey and bear fruiting bodies (acervuli).

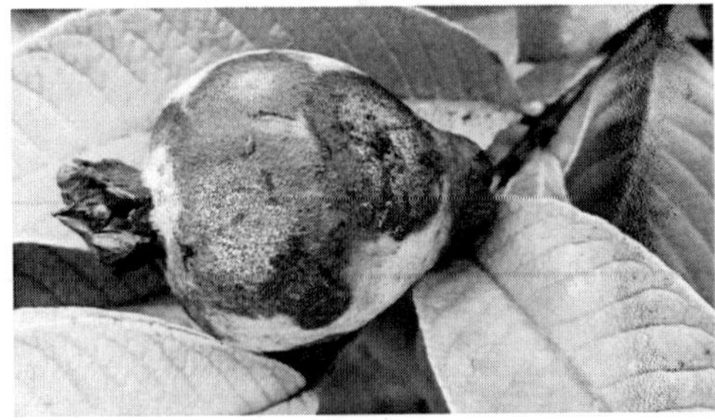

Figure 12.1 Anthracnose of Guava. Sunken lesion with pinkish mycelial growth and acervuli on fruit.

Casual Organism and Etiology

Colletotrichum psidii, Colletotrichum gloeosporioides/Gloeosporium psidii (=*Glomerella psidii/ Colletotrichum psidii*) and *C. gloeosporioides* (teleomorph: *G. cingulata*) are the pathogens responsible for causing anthracnose. The teleomorph stage may or may not play a role in the disease cycle. Production of aerial mycelia by strains varies, ranging from a thick mat to sparse tufts associated with fructifications. Conidia are hyaline, unicellular, and either cylindrical with obscure ends or ellipsoidal with a rounded apex and a narrow, truncate base. They form on hyaline to faintly brown conidiophores in acervuli that are irregular in shape. Setae are 1 to 4 septate, brown, slightly swollen at the base, and tapered at the apex.

Disease Cycle and Epidemiology

Conidia are the most important means by which the pathogen and disease spread in the field. Conidia are produced on necrotic lesions of fruit and inflorescences, mummified fruits, dead twigs, leaves, and other host tissues in the orchard. Conidia are spread by rain splash and cause infection on any aboveground host tissue. A dense canopy is congenial for the germination of spores due to high moisture conditions. Inflorescences and young fruit are extremely susceptible, and if infected, may cause abortion and abscission. Disease spread is greatest during wet conditions. On fruit, lesions develop at any stage of development but expand most rapidly at 30°C. Epidemics can develop during prolonged warm, wet weather, and disease development. Dew or rains encourages spore production and its dispersal around the canopy. Disease is favoured by the temperature range between 10°C–35°C with best 24°C–28°C. Mechanical injuries created during and after harvest can cause the most devastating post-harvest disease.

Management

Cultural Methods

- Orchard sanitation
- Use of micro-irrigation

Host Resistance

- The use of resistant cultivars provides the most efficient tactic in disease management.

Chemical Control

- Spray of Bordeaux mixture (3:3:50) or Copper oxychloride (0.3%) just after initiation of disease.
- For post harvest treatment, dip fruits in 500 ppm Tetracycline solution for 20 min. is effective.
- Apply bioagent, viz. *Streptosporangium pseudovulgare* on fruits before the emergence of symptom.

Banana (*Musa paradisiaca* L.)

13.1 PANAMA WILT

Fusarium wilt or Panama disease is among the most destructive plant diseases of bananas. Fusarium wilt of banana was first discovered in Australia by Bancroft (1876).

Symptoms

The symptoms are most pronounced on 5-month-old plants although 2–3 months old plants are also killed. Fusarium wilt or Panama disease of banana produces two types of external symptoms: *yellow leaf syndrome* and *green leaf syndrome*.

Yellow leaf syndrome: This is the most conspicuous and classic symptom of a Fusarium wilt on the banana. It is characterised by a yellowing of borders of older leaves. The yellowing of the leaves progresses from older to younger leaves (Figure 13.1A). The leaves collapse gradually; bend at the petiole, commonly close to the midrib, and hang down, forming a "skirt" of dead leaves around the pseudostem.

Green leaf syndrome: The leaves of the affected plants eventually stay predominantly green (no chlorosis) until petioles bend and collapse.

In general, younger leaves are the last to show symptoms, frequently remaining unusually erect, giving a bristle-like appearance. Emerging leaf lamina can be markedly reduced, shriveled, and distorted. The most characteristic symptom is seen by cutting through the pseudostem of an affected plant near ground level where a pale yellow, reddish to dark brown and later black discolouration of the water-conducting tissues (xylem vessel) is evident (Figure 13.1B). The youngest infection in banana pseudostems is often yellow to dark red and limited to the xylem vessels only. Pseudostem often shows the more or less conspicuous longitudinal splitting of the outer sheath at the plant base. Longitudinal section through the diseased root base shows characteristic red strands passing into the stele of the rhizome. There is no evidence of symptoms in the fruits.

Figure 13.1 Fusarium Wilt of Banana. (A) Yellowing of leaves, (B) Discoloration of vascular tissues. *Courtesy:* G. Blomme and M. Dita, Biodiversity International.

Causal Organism and Etiology

Fusarium wilt or Panama disease of banana is caused by the fungus *Fusarium oxysporum* f. sp. *cubense*. The mycelium is mainly intracellular (typically found in wood vessel) but it may also be intercellular (in the cortex of roots and parenchymatous tissues). The fungus produces numerous micro-, macroconidia, and chlamydospores. In nature, sporodochia (conidiomata) arise from a globose mass of pseudo-parenchymatous tissue, is produced on the infected surface of petiole and leaves. Microconidia are 0 to 1-septate, ovals to kidney-shaped, and produce in abundance in false heads on branched and unbranched monophialides. Macroconidia are mostly 3-septate, hyaline, sickle-shaped with an attenuated apical cell and foot-shaped basal cell. Chlamydospores are oval to spherical, formed singly or often in pairs. No perfect stage of *F. oxysporum* is known.

Disease Cycle and Epidemiology

The pathogen *Fusarium oxysporum* f. sp. *cubense* is a soilborne fungus surviving on soil as chlamydospores formed by hyphal and conidial cells (macroconidia). The fungus invades the plant mostly through the roots or rhizome wounds. Nematodes help in exposing the roots to infection. It colonizes the vascular tissues and produces masses of mycelium, bearing micro- and macroconidia and chlamydospores. Chlamydospores in plant debris can survive up to 30 years. The fungus also survives as a saprophyte for a long period in the rhizome and other plant parts. After germination, hyphae adhere to and directly penetrate the epidermis; mycelia then advance intracellularly through the cortex and reach the xylem vessels. Once the fungus remains within the xylem, produce mycelium, conidia, and toxins that move upstream in the plant sap, colonizing neighboring vessels and producing new fungal structures. Typical symptoms of wilting are the result of severe water stress due to occlusion of the perforated plates of the xylem vessels as well as by the combination of pathogen activities such as mycelia accumulation, toxin production, and or host defense response including tylose production, gum and vessel shrink due to growth of parenchymatic companion cells. On germination, the chlamydospores produce mycelium on which conidia and chlamydospores are formed. Infection is always through injured secondary or tertiary roots. *Fusarium oxysporum* f. sp. *cubense* can be dispersed through planting material (infected rhizomes or suckers), soil adhering to farm implements and contaminated plant parts, soil, wind, irrigation water, pruning practices and insect vectors, especially the banana weevil borer (*Cosmopolites sordidus*). Usually, the disease is more intense during the warmer and wet months of the year. The disease occurs in warmer parts, where the temperature ranges between 20°C–30°C, and the soil is heavy and acidic.

Management

Cultural Methods

- Use planting material from pathogen-free plantations or preferably from tissue culture.
- Avoid poorly drained soils.
- Follow proper crop rotation.
- Use pathogen-free suckers.
- Practice clean cultivation with proper fertilization, irrigation, and weed control.
- Provide good drainage, especially during the rainy season.
- Use organic soil amendments.
- Flood fallowing has also been found effective method of reclaiming infested soil.

Legislative or Regulatory Measure

- Observe quarantine measures designed to prevent the spread of subtropical and tropical race 4 of Fusarium wilt.

Host Resistance

- Grow wilt-resistant cultivars in endemic areas or infested soil and avoid susceptible varieties.
- Avoid growing susceptible cultivars, viz., Rasthali, Monthan, Red Banana, and Virupakshi.
- Grow resistant cultivar Poovan and tolerant varieties such as Dwarf Cavendish, Robusta, Fhia 1 (Gold finger), Anai komban, and Nivedya Kadali.

Biological Control

- Apply neem cake @ 250 kg/ha.
- Apply bioagent *Trichoderma viride* and *Pseudomonas fluorescens* along with farmyard manure and neem cake.

Chemical Control

- Dip suckers in 1% Carbendazim before planting.
- Drenching of Carbendazim 0.2% solution @ 2–3 litres per plant in the soil around the plants 3–4 times at monthly intervals starting from 1st month of planting.
- Remove and destroy the affected leaves followed by spraying with Bordeaux mixture (1%) + linseed oil (2%).

13.2 BACTERIAL WILT

13.2.1 Banana Xanthomonas Wilt (BXW)

Xanthomonas wilt is commonly known as banana Xanthomonas wilt (BXW), banana bacterial wilt, or enset wilt, is a devastating disease and a major constraint to banana production. The disease can cause up to 100% yield losses if proper management strategies are not well implemented.

Symptoms

The typical symptoms include yellowing, wilting, and necrosis of leaves, buds, and flowers, premature and uneven fruit ripening, internal brown discolourations of fingers and vascular tissues, pale yellow ooze from cut surfaces, wilting of bracts and male buds, and progressive yellowing leading to complete wilting and eventually, the plant dies.

Causal Organism and Etiology

Banana Xanthomonas wilt is caused by the bacterium *Xanthomonas campestris* pv. *musacearum*. The bacterium is a motile, gram-negative, rod-shaped possessing a single polar flagellum. The bacterium produces yellow, circular, mucoid, slimy colonies on nutrient agar and semi-selective medium YTSA-CC.

Disease Cycle and Epidemiology

Xanthomonas campestris pv. *musacearum* survives in infected crop residues in soil and infect the plant through injured roots and rhizome. Nematodes cause root injuries which facilitates the infection. The bacterium is highly transmissible and spreads rapidly through infected plant material, symptomless infected suckers, cutting tools, unintentional moving of infected soils on boots, long-distance trade (of infected fruit, leaves, and suckers), and vectors such as insects, birds, bats and birds sucking nectars or feeding on ripen fruits, grazing animals. Secondary infection occurs through insects which transmits the bacterium from plant to plant. The most common insects are stingless bees (Family *Apidae*), fruit flies (Family *Drosophilidae*), and grass flies (Family *Chloropidae*). Removal of infected suckers creates numerous openings through which yellowish bacterial masses ooze out. This pool of bacteria might act as a source of inoculum and transmited by the insects.

Management

Cultural Methods

- Use healthy pathogen-free suckers for planting.
- Immediately remove and destroy infected plants.
- Control nematode population.
- Remove male bud (debudding) from plants, as the male buds are the main site of infection.
- Disinfect farm implements with formaldehyde diluted with water in 1:3 ratio or disinfected with a 2.5% solution of bactericide Sodium hypochlorite.

Regulatory Measures

- Strict control on the movement of banana planting materials and products.

Chemical Control

- No useful chemical control measures are available for the management of any of these bacterial wilt diseases.
- Apply Copper Oxychloride @ 3 g/L and Streptocycline @ 0.5 g/L as a soil drench.

13.2.2 Ralstonia Bacterial Wilt

Ralstonia Bacterial Wilt is also known as Moko disease. Moko disease was recorded by Schomburgh (1840) in Guyana and the first time described by Rover (1911). In India, Moko disease was reported by Chattopadhyay and Mukhopadhyay in 1968 from West Bengal. In India, crop losses due to Moko disease were up to 70% during 1977–78.

Symptoms

Symptoms of Moko disease include general wilting and subsequent death of the plant, starting with the yellowing then the death of younger leaves and necrosis of the flag leaf or "cigar" leaf. Eventually, disease progress toward the oldest leaves and all other parts of the plant. The wilted plant may fall over. Fruit may show uneven ripening, distortion, splitting, necrosis, a firm brown or grey dry rot of pulp, and bacterial ooze. When fruit, rachis, leaf sheath, pseudostems, or rhizomes are cut, internal damage appears as reddish-brown to black vascular discolouration and rapidly whitish bacterial ooze forms, mainly in the peduncle or pseudostem. The sequence of symptoms depends on the route of infection and the ecotype of the bacterial strain.

- **Infection via the roots and rhizomes:** Yellowing and wilting of the oldest leaves occur first and the plants collapse and bacterial ooze appears on the male inflorescence.

- **Infection via insect:** The symptoms occur initially in the flowers bud and peduncles, which become blackened and shriveled. Then bacteria spread up to fruit and cause internal rot of fruit, later the entire plant is infected and collapses.

The symptoms of Fusarium wilt and bacterial wilt (Moko) of banana may be confusing. See the difference in Fusarium wilt and bacterial wilt (Moko) of banana in Table 13.1.

Table 13.1 Difference in Fusarium wilt and bacterial wilt (Moko) of banana

Features	Panama wilt	Moko disease
Causal organism	*Fusarium oxysporum* f. sp. *cubense*	*Ralstonia solanacearum* race 2
First symptoms	Chlorosis, yellowing, and collapse of the older leaves.	Chlorosis, yellowing, and collapse of the three youngest leaves.
Disease progression	From older to younger leaves.	From younger to older leaves.
Symptoms on young buds or suckers	No wilting symptoms in young growing buds or suckers.	Wilting and blackening of young suckers Young emerging buds can be distorted and necrotic, eventually, die.
Fruit symptoms	No symptoms develop in fruits.	Internal fruit rot and necrosis.
Pseudostem symptoms	Vascular discolouration concentrated near the center	Vascular is discolouration common in the peripheral region.
Production of exudates	No exudations in exposed surfaces.	Bacterial ooze can be observed on exposed cut surfaces (roots, pseudostem, rachis, flowers, rhizome, etc.).

Causal Organism and Etiology

Ralstonia solanacearum (*Ralstonia solanacearum* race 2, biovar 1 strains). The bacterium is described as an aerobic, gram-negative, non-fluorescent rod belonging to the rRNA homology group II.

Disease Cycle and Epidemiology

The bacterium survives in infected plant material, vegetative propagative organs, wild host plants, soil, and asymptomatic weed hosts associated with banana cropping. Ralstonia wilt can be spread through contaminated plant sources, root contacts, sick soil, water, insects (including common pollinators, and banana pests), cultivation, and harvesting equipment. Primary infection occurs when the bacteria enter the roots and rhizomes through wounds, caused by nematodes, insects, and farm implements. Secondary infection is caused by insect, particularly stingless bees (*Trigona* spp.) that transmit the bacterial oozes from infected male flowers bracts to healthy flowers. Generally, high temperatures and high soil moisture favour disease development.

Management

Cultural Methods

- Use certified seed.
- Provide good field drainage.
- Remove infected crop debris and infected plants and destroy them.

- Manage nematode population.
- Rotate crop for 3 years rotation with non-solanaceous crops such as rice, corn, bean, cabbage, and sugarcane.
- Disinfect farm implements with Formaldehyde diluted with water in 1:3 ratio or disinfects with a 2.5% solution of bactericide Sodium hypochlorite.
- Allow fallow period or flooding during the off-season.
- Follow the practice of removing the male bud from the male axis as soon as the last female hands emerged.
- Avoid water stress. Plants are more susceptible to rhizome rots when water-stressed in hot and dry conditions followed by heavy rainfall.

Biological Control

- Lytic bacteriophages can also be used to manage Moko disease (*Ralstonia solanacearum*) of banana.
- Incorporate French marigold (*Tagetes patula*) into the soil to reduce *R. solanacearum* populations.
- Use bio-agent like *Pseudomonas fluorescens*.

Chemical Control

- No useful chemical control measures are available for the management of any of these bacterial wilt diseases.
- Apply Copper Oxychloride @ 3 g/L and Streptocycline @ 0.5 g/L as a soil drench.

13.3 SIGATOKA LEAF SPOT

Sigatoka leaf spot leads to 11%–80% yield losses in banana by reducing the photosynthetic tissues through necrotic leaf lesions. Two types of leaf spots are produced in the Sigatoka leaf spot disease of banana: yellow and black. Black Sigatoka was first recognised in the Sigatoka Valley of Fiji in 1963, now it is widespread in Southeast Asia and the South Pacific. Black Sigatoka causes significant yield losses upto 50% and is more damaging and difficult to manage than yellow Sigatoka disease. Yellow Sigatoka was the first leaf spot disease to have a global impact on bananas. Yellow Sigatoka was named after an outbreak of the disease in the Sigatoka Valley on the island of Viti Levu in Fiji in 1912. Among the landraces, French plantain cv. Nendran (AAB), grown extensively in the southern most states of India and especially in Kerala, is highly prone to this disease.

Symptoms

Yellow Sigatoka: Symptoms start as minute yellowish-green specks, 1–2 mm long that grow into long yellow streaks. Later these streaks broaden and form large spots with the sunken center, yellow hallow, and dark brown margins (Figure 13.2). The leaf tissues die rapidly and large areas of the leaf turn brown and grey. The most distinguishing features of yellow sigatoka disease are yellow rings around the young spots and yellowing of leaves around the mature spots.

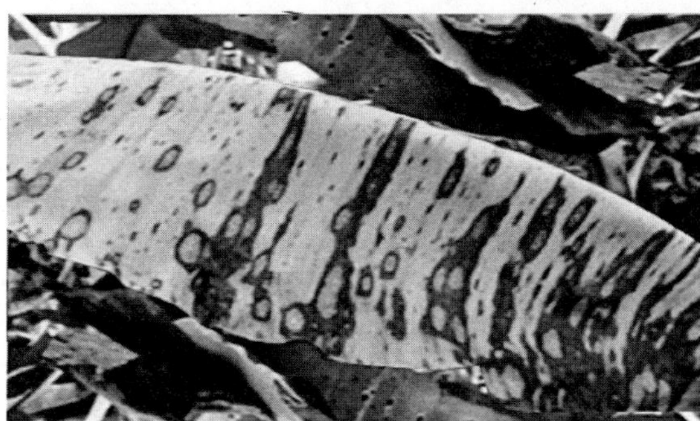

Figure 13.2 Yellow Sigatoka of Banana. *Source:* Tamil Nadu Agricultural University, India (Agritech Portal).

Black Sigatoka: Symptoms are similar to yellow sigatoka, except that the streaks are rusty red to brown, which later become black (Table 13.2). The first symptoms appear as tiny, chlorotic spots or specks, 1–2 mm long, on the lower surface of the leaf (Figure 13.3). Gradually, these spots expand, widen, and darken to form necrotic rusty-reddish brown streaks that are limited by leaf veins. Eventually, these lesions then enlarge (20–30 mm long), and become oval-shaped or fusiform or elliptical with sunken, grey, or light brown center and dark brown or black margins. The streaks broaden and become visible on both leaf surfaces. Large areas of the heavily infected leaves may become blackened and water-soaked hence named "black sigatoka disease". Numerous, tiny, black, globose fruiting bodies (pseudothecia), emerge from the necrotic tissues of the underside of the leaf.

Table 13.2 Difference in yellow and black sigatoka spots of banana

Features	Yellow Sigatoka	Black Sigatoka
Causal organism	*Mycosphaerella musicola* (*Pseudocercospora musae*)	*Mycosphaerella fijiensis* (*Pseudocercospora fijiensis*)
Characteristics of conidia and conidiophores	Conidia are uniform thickness for full length, 1–5 septate; no distinct basal scar; Conidiophores are usually short and bottle-shaped.	Conidia taper from base to apex, 1–6 septate, distinct basal scar; Conidiophores are elongated, often bent, and have conspicuous conidial scars.
Early symptoms	Streaks are pale yellow or yellow green, narrower, and shorter and more prominent on the upper leaf surface	Streaks are reddish to rusty-brown, longer and broader, and more prominent on the lower leaf surface.
Host range	Bananas (AAA) are generally susceptible; most cooking bananas and plantains (AAB and ABB) are moderately to highly resistant.	Most dessert bananas, cooking bananas, and plantains are susceptible.
Sporulation	Produce more than 30,000 conidia per spot.	Produce about 1200 conidia per spot.
Germination of spores	Requires a film of free water for conidial germination and 95% relative humidity for ascospore germination.	Requires high relative humidity (92–100%) for conidial germination and free water for ascospore germination.
Conducive environment for disease development	More common in cooler environments.	More common in warmer environments.

Figure 13.3 Black Sigatoka of Banana.

Causal Organism and Etiology

Sigatoka leaf spot caused by the ascomycete fungal complex *Mycosphaerella* spp. Sigatoka leaf spot diseases, viz., yellow sigatoka (*Mycosphaerella musicola*, anamorph *Pseudocercospora musae*), black Sigatoka (*Mycosphaerella fijiensis,* anamorph *Pseudocercospora fijiensis*) and Eumusae leaf spot (*Mycosphaerella eumusae*, anamorph *Pseudocercospora eumusae*). Among these, yellow and black sigatoka is considered the major leaf spot diseases of banana. The pathogens produce conidia and ascospores, both of which are infective. Conidia and ascospores are spread by wind, and in the case of conidia, also by rain and irrigation water. Ascospores are more important than conidia (due to their greater abundance and small size) in spreading the disease within plants and plantations.

Disease Cycle and Epidemiology

The disease cycles of *M. musicola* and *M. fijiensis* are essentially the same and consist of four distinct stages: survival, spore germination, host penetration and infection, symptom development and reproduction, and spore dispersal.

Survival: Both fungi survive in infected plant parts and crop residues which serve as a source of primary infection.

Spore germination: The disease cycle begins when conidia or ascospores germinate on the leaf surface. Germination is highly dependent on moisture and temperature; stomatal penetration by germ tubes occurs when temperatures are above 20°C for 2–3 days and moisture is nearly 100% relative humidity.

Host penetration and infection: After germination, both species undergo a period of epiphytic growth. In *M. musicola*, this is 2–3 days for ascospore germ tubes and 4–6 days for conidial germ tubes. Subsequently, the germ tube produces an appressorium over a stomatal pore through which the fungus directs a fine, infection hypha. The same infection process has been observed for *M. fijiensis* and *M. musicola.*

Symptom development: In general, the first symptom appears as small, pale yellow streaks (*M. musicola*) on the upper leaf surface or small reddish-brown streaks on the lower leaf surface (*M. fijiensis*), both 1–2 mm long, which enlarge to form necrotic lesions with yellow haloes and light grey centers.

Reproduction and spore dispersal: Both fungal species are capable of asexual and sexual reproduction, producing conidia in sporodochia and ascospores in perithecia. Sporodochia are produced on both leaf surfaces within the substomatal air chambers, and the conidiophores grow out through the stomatal pore. *M. musicola* produces more than 30,000 conidia per spot while *M. fijiensis* produces about 1200 conidia per spot. Both conidia and ascospores which are formed under high moisture conditions are disseminated by the wind. The conidia are also spread by rain and irrigation water. Conidia disperse during rain-wash and splashing, causing the local spread of the disease. The ascospores spread disease to long distances as they are carried upwards by air currents and are smaller and more abundant. The infected planting material and leaves, which are used often in the developing world as packing materials, are usually responsible for the long-distance spread of the disease. Ascospores are the primary means of long-distance dispersal and are the main means of spreading during extended periods of wet weather. *Mycosphaerella fijiensis* forms relatively few conidia, so ascospores are thought to be more important in the disease cycle. Both conidia and ascospores are infective. The spores germinate and the germ tubes enter through natural openings in the leaf. The fungus grows within the leaf, killing plant cells, before returning to the surface to produce more spores.

Favorable environmental conditions: With black Sigatoka, ascospores, and to a certain extent conidia, are the propagules by which the fungus is dispersed readily in high humidity, especially if a film of free water is present on leaves. The conidia disperse during rain-wash and splashing, causing local spread of the disease. Pseudothecia mature when dead leaf tissues are saturated with water for approximately 48 hours. Sigatoka leaf spots decrease somewhat during the dry season but otherwise produce more or less continuously repeated cycles of infection. Conidia germinate during periods of high relative humidity (92%–100% relative humidity) and infect the leaf through a stoma, usually on the underside of a leaf.

Management

Cultural Methods

- Remove infected leaves or part of leaves and destroy them outside the orchard.
- No dried leaves should be hanging around the plant.
- The field must be kept weed free and clean. Follow either hand weeding/harrowing till 5 months after planting or use herbicide glyphosate 7–10 ml/litre of water + 25 g of urea or ammonium sulphate/tank or by intercropping with cowpea.
- While planting, optimum/recommended spacing (1.6 m × 1.6 m) must be followed.
- Provide adequate drainage facility whenever it is required.
- Apply only the recommended dose of fertilizer: N, P, K g/plant (200:40:400) as per the schedule + 25 g Azospirilium + 25 g phosphorus solubilizing bacteria. Potash can be applied 10 to 20% more. Micronutrient mixture 10 g/plant in 3rd month and 5th month after planting must be applied.

Biological Control

- Apply Neem cake @ 0.5 to 1 kg/plant.

Chemical Control

The following pesticides may be applied as soon as the symptom appears on the leaves. The interval between two sprays may be 20 to 25 days. The fungicide requirement for one litre of water is as follows:

(a) Mineral oil (10 ml) + Propiconazole (0.1% or 1 ml per litre of water).
(b) Mineral oil (10 ml) + Carbendazim (0.1% or 1 g per litre of water).
(c) Mineral oil (10 ml) + Tridemorph (0.1% or 1 g per litre of water).
(d) Mineral oil (10 ml) + Mancozeb + Carbendazim (0.1% or 1 g per litre of water).
(e) Mineral oil (10 ml) + Propiconazole (0.1% or 1 ml per litre of water).
(f) Mineral oil (10 ml) + Carbendazim (0.1% or 1 g per litre of water).
(g) Mineral oil (10 ml) + Tridemorph (0.1% or 1 ml per litre of water).
(h) Mineral oil (10 ml) + Propiconazole (0.1% or 1 ml per litre of water).

Notes:

1. No sticking agent is required when oil (Banole–banana spray oil) is used.
2. Spray must be done in such a way that the spray should cover both sides of the leaves and all the leaves including the top most unfurled leaf.
3. Apply only the recommended dose of chemicals to avoid the development of resistance to fungicides, the cost of cultivation, to reduce environmental pollution, and also to avoid residue problems in the production.

13.4 BANANA BUNCHY TOP DISEASE

Banana bunchy top disease (BBTD) is the most serious virus disease of bananas and plantains worldwide. BBTD was first reported in Fiji in 1889 and caused heavy destruction that threatened Fiji's banana export industry. BBTD was introduced in Sri Lanka in 1913. Later, from Sri Lanka, it is introduced in India in 1940 and eventually spread to various banana-growing areas in the entire country. The disease is endemic in many banana-producing countries in Asia, the Pacific, and Africa. In the 1970s, the virus has been the main cause of drastic reduction in hill banana cultivation and showed the incidence of 14%–74%. Bunchy Top Virus (BBTV) is listed among the world's worst 100 invasive species.

Symptoms

Characteristic severe symptoms include intermittent dark green dots and streaks, so-called "Morse-Code" patterns, of variable lengths (1 to 1.25 mm long and 0.75 mm wide) in leaf sheath, midrib, leaf veins, and petioles of infected plants. Sometimes, the typical streaks may not appear. The leaves produced are progressively shorter, brittle, narrow, more upright clusters, leaf edges often roll upwards with pale yellow and wavy margins and give the "rosette" or 'bunchy top' appearance of newly emerging leaves, at the apex of the plant. The affected plant fails to produce a fruiting bunch or produces a bunch with fingers that do not develop further and reach maturity. When infection takes place very late in the season, no leaf symptoms may appear, but dark-green streaks/flecks may be seen on the tips of the bracts. Infected plants do not produce fruit: bunches either fail to form or fail to emerge from the pseudostem depending on the time of infection, resulting in significant yield losses.

Casual Agent and Etiology

Banana bunchy top disease is caused by *Banana bunchy top virus* (BBTV) belonging to the genus *Babuvirus* (family *Nanoviridae*). The virus particle is isometric, 18–20 nm in diameter, with a multicomponent genome comprising at least six circular, single-stranded DNA molecules. Banana is the only host of this virus.

Disease Cycle and Epidemiology

Banana bunchy top virus is primarily disseminated through vegetative propagules, including suckers, corms, and tissue-cultured plants. All suckers produced by a diseased plant carry the virus and constitute the most important source of the spread of the disease from plantation to plantation. The virus is transmitted by the banana aphid, *Pentalonia nigronervosa* in a persistent, circulative, non-propagative manner, and the vector has worldwide distribution. It is not a sap-transmitted virus. The aphid causes secondary spread during plant growth. The multiplication rate of the aphid is maximum when the minimum temperature is about 18°C–20°C with relative humidity ranging from 41%–84%. All stages of the aphid can acquire and transmit the virus but nymphs are the most efficient. The aphid requires a minimum feeding period of 4–18 hours on phloem, to be infective and transmits the virus on susceptible plants in 30 minutes to 2-hour feeding period. Aphids can retain their infecting capacity for a period of 13 days and may cover large distances, especially when blown by the wind. There is no transovarial transfer, i.e., the virus does not transmit from adult aphids to their offspring. Banana aphids are all female and give birth to live young. They complete their life-cycle from nymphs to adults in 9–16 days.

Management

Cultural Methods

- Use of virus-free certified planting materials.
- Destroy all volunteer plants.
- Remove and destroy crop residue and infected banana plants with all their attached suckers.
- Practice clean cultivation.

Host Resistance

- Grow resistant cultivars. Some of the susceptible varieties are Virupakshi (Hill banana), Grand Naine, and *Musa* species like *M. flaviflora* and *M. burmannicoides*.

Biological Control

- Biocontrol of the aphid can be done through the parasitic wasp, *Aphidus coleman*.

Chemical Control

- Inject 4 ml of Fernoxone solution (50 g in 400 ml of water) or insert Fernoxone capsules (containing 200–400 mg of chemical per capsule) into the pseudostem to kill the virus infected plants.
- Spray Dimethoate 30 EC @ 1.0 ml/L of water along with stickers to control insect vectors.
- Spray the infected plant immediately with kerosene or mineral oil to kill all aphids.

14 Papaya (*Carica papaya* L.)

14.1 STEM OR FOOT ROT

The disease is also known as collar rot or root rot. This is the most important and serious disease of papaya and is widespread in India.

Symptoms

Stem or root rot: Typical stem rot is most common in 2–3 years old plants. Initially, patches of water-soaked lesions appear on the stem at the soil line. These lesions enlarge and girdle the entire base of the stem and finally rot and appear dark brown to black. The terminal leaves droop, wilt, and fall off prematurely. Sometimes, only a few small leaves remain at the top of the tree. If fruits are produced, they are also infected. Fruits rot quickly and are covered with white mycelial growth containing spores of the water molds and later fall. The affected plants topple over and die within a short period due to rotting and disintegration of the parenchymatous tissue of the stem and roots. The tissues below the bark give a "honeycomb appearance", if bark is opened. The plants can die at any stage of their growth.

Damping off: In nurseries, damping off is also very common and caused by the same pathogen.

Causal Organism and Etiology

Pythium species (Oomycetes) are reported to cause stem rot of papaya in India and *Pythium aphanidermatum* is responsible mainly for stem rot and damping off in Hawaii. *Phytophthora palmivora* and *Phytophthora nicotianae* has also been reported to be the cause of this disease in Australia and Fiji. In addition, they cause fruit rot.

Disease Cycle and Epidemiology

The pathogens are soilborne. The oospores are formed on the infected papaya residue in the soil. The pathogen can also survive as a saprophyte on dead organic matter. The secondary spread takes place by zoospores. Water molds survive as thick-walled resting spores called "chlamydospores" in the soil. Under favourable conditions, the chlamydospores germinate and produce sporangia. Inside the sporangia "zoospores" are produced, and these are capable of swimming short distances in the water between soil particles. Zoospores germinate and infect the fine feeder roots of papaya. Spores of water molds are spread through rain splash, wind-driven rain, irrigation water and in soil on machinery and shoes. Environmental conditions conducive to disease development are optimum temperature 36°C and abundance moisture around the base of the stem.

Management

Cultural Methods

- Use disease-free seedlings for planting.
- Sow 4–5 seeds per bag then retain only 3 seedlings.
- Use two-month-old seedlings for transplanting.
- Transplant seedlings or establish an orchard in well-drained soil, where water logging does not occur.
- Uprooted and burnt infected plants.
- Avoid planting in the same place or pit.

Chemical Control

- Treat papaya seeds with Thiram or Difolatan @ 0.25% for improving seedling emergence.
- Apply the fungicidal paste at the early stages of infection after removing infected tissues.
- Soil drenching and spraying the stem with 6:6:50 Bordeaux mixture or Captan @ 0.2% may reduce the incidence of the disease of stem rot.
- Treat the soil (soil drenching) with Captan @ 0.2% before and after seedling emergence to avoid the damping off phase.
- For managing foot rot caused by *Phytophthora* spp. apply Metalaxyl + Mancozeb @ 1% in the soil during transplanting followed by weekly sprays of Potassium phosphonate @ 1 g/L + Acibenzolar-S-methyl @ 0.025 g/L.
- Application of Benzothiazole (BTH) has been shown to induce systemic acquired resistance (SAR) in papaya against *Phytophthora* spp.

14.2 VIRAL DISEASES OF PAPAYA

14.2.1 Papaya Leaf Curl

The disease is common in papaya in India but the incidence is less as compared to PRSV. The affected plants do not bear any fruit causing heavy loss to the farmers. Papaya leaf curl disease was first reported in 1939 by Thomas and Krishnaswamy. This disease is of moderate incidence and widely distributed in India. *Papaya leaf curl virus* is a major threat for the crop production, and the virus has the capability to adapt to new plant hosts very rapidly which helps in their host range extension that also has emerged as an evolving risk in papaya production.

Causal Organism and Etiology

Leaf curl of papaya is caused by *Papaya leaf curl virus.* The causal virus particles are geminate and the virus belongs to family Geminiviridae. The genome of the virus is ssDNA and divides into two parts, DNA-A and DNA-B.

Symptoms

The most characteristic symptoms are severe curling, crinkling, and distortion of leaves accompanied by vein clearing and reduction of leaf size (Figure 14.1). Leaves become leathery, brittle and the interveinal areas become raised due to hypertrophy which develops rugosity. The most prominent symptoms are rolling of leaves downward and inward in the form of an inverted cup and thickening of the vein. The petioles of leaves are twisted in a zig-zag manner. The growth of affected plants is reduced and affected leaves appear as a bunch of leaves at the top. Affected plants do not bear any flowers and fruits. If fruit set occurs, then diseased plants may bear a few small distorted fruits and tend to fall prematurely. In the advanced stages of the disease, the plant defoliate and growth is arrested.

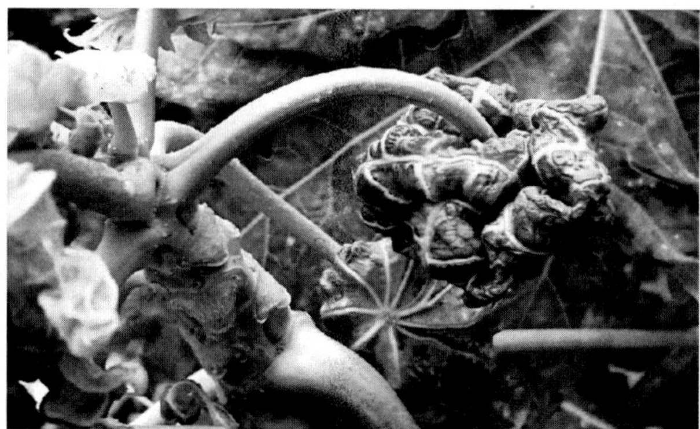

Figure 14.1 Papaya Leaf Curl.

Disease Cycle and Epidemiology

The inoculum survives on papaya in nature or may be on other plants like tobacco, tomato, and several weed hosts. From these sources, the virus infects papaya with the whitefly vector. The virus is transmitted by *B. tabaci* in a persistent, circulative manner. The virus is not transmitted through sap or mechanically and seed but can be by grafting. Affected plants never recover from the disease and hence are a perennial source of inoculum.

Management

Cultural Methods

- Keep papaya field weed free.
- Raise papaya nursery in the net house or in whitefly-free isolated places and if needed spray Imidacloprid @ 0.03% regularly at 21 days interval.
- Uproot and destroy the virus-affected plants. As removal and destruction of the affected plants is the only control measure to reduce the spread of the disease.
- Avoid growing tomato, tobacco, sunnhemp, petunia, chilli, cape gooseberry, *Datura stramonium*, *Zinnia elegans*, etc. nearby papaya field.
- Grow border or intercrop such as sorghum, maize, and pearl millet to reduce whitefly infestation.

Host Resistance

- No source of resistance is available.

Biological Control

- Spray regularly neem-based biopesticides such as Nemaban or Neemagan @ 0.3% at 15 days intervals.

Chemical Control

- Manage vector (whitefly) population. The vector may be kept under control by 4–5 sprays of Diamethoate (@ 0.05%), or Metasystox (@ 0.02%), or Nuvacron (@ 0.05%).

14.2.2 Papaya Mosaic

The mosaic disease is very destructive; losses ranging from 5%–20% are common in many orchards but losses as high as 75% have occurred. *Papaya mosaic virus* (PapMV) was first reported in Florida (USA) in 1962. Papaya mosaic was first described by Conover (1964). Capoor and Verma (1958) reported this disease from Bombay. The disease is widespread in India.

Symptoms

- **On leaves:** Mild mosaic or mottling, vein clearing, puckering, rugose and abnormal pattern, most evident in young leaves.
- **On whole plants:** Dwarfing or stunting of plants. No symptom appears on the stem, petioles, and fruits.

Causal Organism and Etiology

Papaya mosaic is caused by *Papaya mosaic potex virus*-1 (PaPMV-1), a member of the *Potexvirus* genus and the family Alphaflexiviridae. It has monopartite, filamentous, flexuous virions, 470–580 nm long and 13 nm wide, and a single-stranded, positive-sense RNA genome of 6.65 kb size.

Host range: Restricted to papaya and cucurbits.

Disease Cycle and Epidemiology

PaPMV-1 is readily sap transmissible and can be spread via plant-to-plant contact. Insect vectors aphid (*Myzus persecae*), transfers the virus and there are no reports of seed transmission. Papaya mosaic disease is mechanically transmissible virus associated with other viral diseases, from *Papaya mosaic virus* being aphid-borne and restricted in host range to papaya and cucurbits.

Management

Cultural Methods

- Plant virus-free seedlings.
- Transplant virus-free seedlings into new field soil or rotate with non-host crops.
- Inactivate the virus on contaminated tools and pots by first removing excess dirt and plant debris, and then, either heating for 1 hour in an oven at 150°C, or soaking for a few minutes in 0.5% Sodium hypochlorite solution (rinse with water to remove excess Sodium hypochlorite).
- When working with plants it is very important to keep hands clean and wear plastic gloves that can be dipped in 0.5% Sodium hypochlorite which is then rinsed in water.
- Since the virus is easily spread by contact, care should be taken to prevent transmission in the nursery and field with contaminated implements and handling of plants.

Mechanical Control

- Remove and destroy by burning virus-infected seedlings per plant.

Host Resistance

- Use resistant cultivars.

Biological Control

- Apply 5% neem seed kernel extract (NSKE) or groundnut oil @ 1% to 2% onto the plants to manage the vector population.

Chemical Control

- Application of Carbofuran 1.0 kg a.i. per hectare in nursery bed at the time of seed sowing followed by 2–3 sprays of Phosphamidon (0.05%).
- Use groundnut oil-fatty acids present in oil not only kill PaPMV but also aphid vectors that transmits and further spread the disease.

14.2.3 Papaya Ringspot

Papaya ringspot was first described by Jensen (1949). PRSV was first detected in Mexico in 1975, where it caused severe damage in the main papaya production states with crop losses of up to 85%. The loss is 100% if the plants are infected in the early stage of their growth. There are some areas in the world where papaya is either withdrawn from cultivation or shifted due to this disease.

Symptoms

Leaves: Mottling, mosaic, dark-green blisters, malformation, yellowing, and vein clearing of young leaves. Leaves often become filiform, looking like "shoe-strings" (tendril-like) due to extreme reduction of leaf lamina (Figure 14.2A and B).

Fruits: Distinctive, small green, ringspot patterns (concentric rings and spots, or C-shaped markings) on immature fruits (Figure 14.2C). The spots are initially about 1 mm in diameter with dark green outer rims and tan centers. As the fruit matures the rings and spots grow larger (4–8 mm in diameter) and become darker in colour, and as the fruit ripens they can change in colour to yellow and brown.

Stems: Translucent water-soaked patches or oily-spots and oily streaks on green stems and petioles.

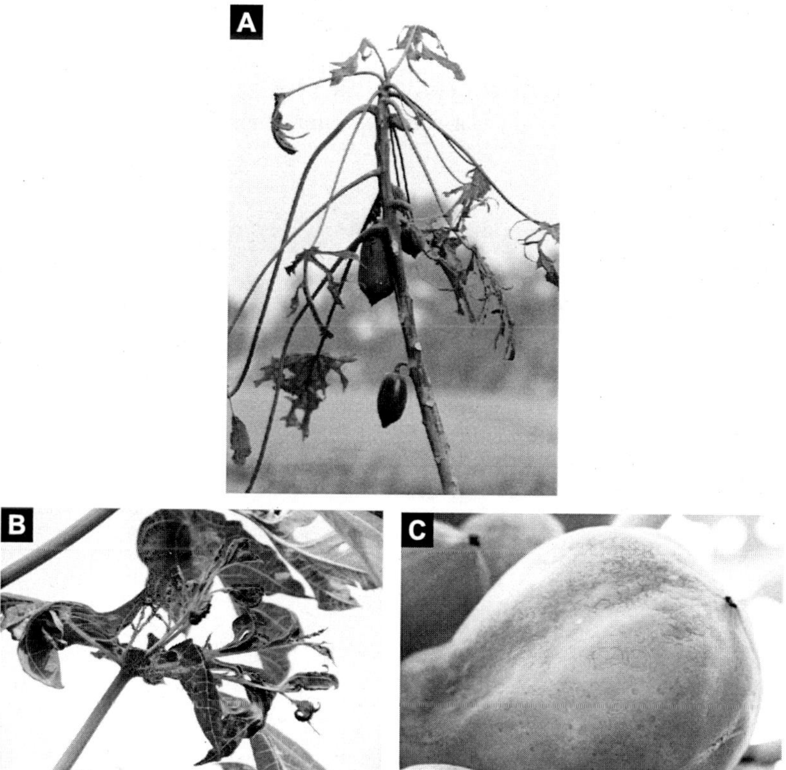

Figure 14.2 Papaya Ring Spot. (A) and (B) Shoestring formation of leaves, (C) Circular ringspots or C-shaped markings on fruit.

Causal Organism and Etiology

Papaya ring spot disease is caused by *Papaya ring spot virus.* The virus belongs to the potyvirus (potyviridae) group. The virions are typically long (760–800 × 12 nm), flexuous, non-enveloped, and filamentous. The genome of the virus is unipartite, linear, and (+)-ssRNA, 12 kb in size.

PRSV has a single type of coat protein of 36 kDa and induces cylindrical (pinwheel) and amorphous inclusions body in the cytoplasm of host cells. The virus has two strains; PRSV-W which infects watermelon but not papaya and PRSV-P strain which infects both papaya and cucurbits. A severe PRSV isolate from Taiwan is also known to induce systemic necrosis and wilting along with mosaic and chlorosis.

Host Range: Cucurbits and papaya.

Disease Cycle and Epidemiology

The virus survives on papaya plants in the field from where it infects healthy papaya by aphid vectors. The amount of primary infection increases with the nearness of infected orchards. The virus is not transmitted through seeds. The virus is also transmitted by mechanical means, like pruning and grafting. Aphids can spread the virus between plants while feeding for less than a minute. PRSV is transmitted by aphid vectors, *Myzus persicae, A. craccivora, Aphis gossypii* etc. in a non-persistent manner, and also spreads to cucurbits. *Myzus persicae* is the most efficient vector. In the tropics and subtropics where both biotypes and their hosts are present, PRSV-P effectively completes its life cycle in papaya while PRSV-W completes its life cycle in cucurbits. In other words, cucurbits generally do not serve as alternative hosts for PRSV-P.

Management

Cultural Methods

- Use disease-free seedlings for planting.
- Raise papaya seedlings under insect-proof conditions.
- Grow sorghum/maize as a barrier crop before planting papaya.
- Rogue out affected plants immediately upon noticing the symptoms.
- Do not raise cucurbits around the field.
- Avoiding vectors and removing the source of inoculum are the best measures for the management of the disease.
- Use of yellow sticky trap for the control of aphid vector.

Host Resistance

- No variety of papaya is resistant.
- Use virus-resistant transgenic like "SunUp" and "Rainbow" which are resistant to all isolates of PRSV-P.
- **Cross Protection:** It is a technique that has been used for the control of papaya ring spots in different countries with varying degrees of success. This technique involves the inoculation of papaya seedlings prior to planting with a mild strain (Nitrous acid mutant of PRS-P) which protects plants against damage caused by infection with severe a strains of the virus in the field.

Chemical Control

- Chemical control of vector is similar to papaya mosaic disease.

15 Pomegranate (*Punica granatum* L.)

15.1 BACTERIAL BLIGHT

In India, the bacterial blight of pomegranate was first reported by Hingorani and Mehta (1952). Bacterial blight caused is a major disease of pomegranate. Bacterial blight drastically reduces the yield and quality of fruits, which are critical for pomegranate production. Severe disease outbreaks can cause 60%–80% yield losses. It has become an increasingly serious threat to pomegranate growers in the states of Karnataka, Maharashtra, and Andhra Pradesh in India. In some orchards of Solapur District (Maharashtra), blight caused a loss of up to 80.0%.

Symptoms

The disease affects all plant parts including leaves, twigs, stems, branches, trunks, and flower, but is most destructive on fruits.

Leaf: Initially, minute (1–5 mm), circular to irregular, water-soaked lesions with necrotic center which later turns light to dark brown surrounded by prominent water-soaked margins or translucent yellow halo (Figure 15.1A). The individual spots of the leaves coalesce and appear as large blighted areas with dried silvery bacterial ooze. Blighted leaves often turn yellow and fall off prematurely.

Stem: Brown to black lesions observed on twigs and stems which extend and often girdle around the nodes of the stems, resulting in cracking of nodes and breakdown of branches. As infections are quite frequent at nodes of the stem, the disease is also known as "nodal blight".

Flower: At the base of the flower, water-soaked lesions appear which gradually turn necrotic black or brown. The infected flowers do not bear fruits and dry up permanently.

Fruit: Fruits are the most susceptible part of the plant to bacterial infections. Blight lesions on the fruits pericarp appear as brown to black, necrotic, small to large L- and Y-shaped fissures (cracks), splitting the entire fruit, and dried thin white encrustation of bacterial ooze (Figure 15.1B). Blight infections on fruits are restricted to the rind (pericarp) portion only. The cracking on fruit later predisposes to the secondary infection by rot causing saprophytic of fungi and bacteria.

Figure 15.1 Bacterial Blight of Pomegranate. (A) Irregular to circular, water-soaked translucent, black spots on the leaves, (B) Black spots on the fruits.

Causal Organism and Etiology

Bacterial blight of pomegranate is caused by *Xanthomonas axonopodis* pv. *punicae*. The bacterium is a non-spore former, gram-negative, rod shaped, motile with a single polar flagellum, produces water-insoluble, nondiffusible yellow pigment (xanthomonadins), and measures 0.4–0.75 × 1.0–3.0 μm.

Disease Cycle and Epidemiology

The bacterium survives in infected plant stems, diseased tissues, buds, fallen leaves, and plant debris in the soil for up to one year. Bacteria on the leaf surface act as secondary inoculum that rapidly spreads through rain splashes, irrigation water, pruning tools, insect vectors (e.g.

butterflies, aphids, and larvae of fruit borer), and human beings. The bacterium enters and infects different plant parts through natural openings like stomata, lenticels, hydathodes, or wounds. The pathogen also spreads via air and enters the host through natural openings or physical injuries to plants. The disease spread to a new orchard through infected planting material and contaminated irrigation water coming from the infected orchard. High temperature (25°C–35°C), relative humidity above 30%, rain, and wind, is highly conducive to rapid disease development.

Management

Cultural Methods

- Plant new orchards with disease-free planting material.
- Avoiding flowering and fruiting during excessive rainy periods; complete defoliation after harvest of the crop followed by 3–4 months of the rest period and maintaining orchard sanitation standards.
- Disinfect farm tools after use with water or bleaching solution or sterilize with flame.
- Do not allow contaminated water from an infected orchard to enter in the healthy orchard.
- Wash the sole of shoes before moving from a diseased orchard to a healthy orchard.

Biological Control

- Spray bacterial antagonistic such as *Pseudomonas fluorescens* and *Bacillus subtilis*, *Streptomyces violaceusnige*, and plant extracts from pongamia, neem, coleus, and periwinkle.

Chemical Control

- Spray a mixture of Streptocycline (@ 200 ppm to 250 ppm) + Carbendazim (@ 0.2%).
- Spray plant growth regulators Ethylene (200 ppm) and Nanocopper (2 ppm).
- Spray Bordeaux mixture (0.5% but 1% just after pruning and rest period).
- Spray Streptocycline (Streptomycin sulphate, 500 ppm) in combination with Copper oxychloride (0.2%) followed by Bronopol (2-bromo-2-nitropropane-1, 3-diol, 500 ppm) and Copper oxychloride (0.2%) and spreader sticker (0.5 ml/L) at 10–15 days interval depending on weather conditions.

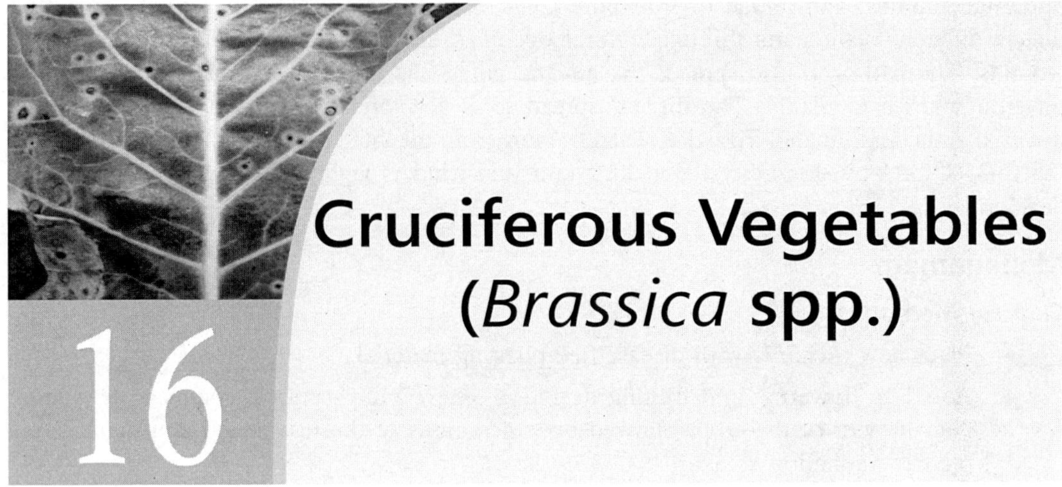

16 Cruciferous Vegetables (*Brassica* spp.)

16.1 ALTERNARIA LEAF SPOT

Alternaria leaf spot of cruciferous vegetables is also known as "head rot" or "dark leaf spot". *Alternaria* pathogens are important on crucifer crops worldwide, causing foliar, pod, seed, and broccoli-head diseases. All the major brassica types (*B. campestris*, *B. oleracea*, *B. napus*, *B. rapa*) are susceptible. Chinese cabbage, bok choy, and leafy brassicas can be seriously damaged and rendered unmarketable. The disease may cause qualitative and quantitative looss upto 10% to 100%. Alternaria pathogens usually cause damping-off of seedlings, spotting of cabbage leaves, blackleg of heads of cabbages, knol-khol, radish, and spotting of cauliflower curds and broccoli florets.

Symptoms

On leaves, initially small, dark-brown to black lesions (1–2 mm in diameter), each surrounded by a halo of chlorotic tissues and mostly in between the veins, develop (Figure 16.1). Later these enlarge (5–25 mm in diameter), become circular, tan to brown, usually contain concentric rings, giving a typical "target board-like appearance" and are visible on both surfaces of the leaves. A zone of the yellow halo surrounding the lesions is caused by a fungal toxin. Dark green mass of conidia is produced on these spots. In general, *A. brassicae* produces grey lesions compared to the black sooty velvety lesions with a more irregular margin by *A. brassicicola.* Older leaf spots become papery and the centers drop out and give a "shot-hole" appearance. As the disease progress, leaf spots coalesce, leaves dry out, and look scorched (blighting) and defoliate. The lesions on the stem and petiole are brown to black, oval or elliptical or elongated with pointed ends. Siliquae may show sunken, dark-brown to black circular lesions. Inside the pods, infected seeds just beneath the black spot are small, shriveled, grey to brown, and covered with fungal mycelial growth. Seedlings produced from infected seed show pre- and post-emergence damping-off due to cotyledons and hypocotyls infection. Dark brown to black, water-soaked decay appears on the heads of broccoli and cauliflower. The head tissues become soft and mushy and the pathogen sporulates. Older leaves and pods are more susceptible to infection and show increased disease as they senesce.

Figure 16.1 Alternaria Blight of Cabbage.

Causal Organism and Etiology

Alternaria leaf spots of *Brassica* spp. are caused by *Alternaria brassicae*, *A. brassicicola*, and *A. raphani*. Among the three spcies *Alternaria brassicae* is more damaging. Two species *A. brassicae* and *A. brassicicola* attack all plant species mentioned above except radish which is attacked by *A. raphani*. See Table 16.1 below for the difference in symptoms and morphological features of these three *Alternaria* species.

Table 16.1 The difference in *Alternaria* species (*A. brassicae*, *A. brassicicola*, and *A. raphani*)

Features	*Alternaria brassicae*	*A. brassicicola*	*A. raphani*
Symptoms	Spots are circular, grey to dark brown almost black, dense, and sparsely covered with brown conidiophores and conidia.	Spots are circular with a more irregular margin, sooty black, velvety and copiously covered with black conidiophores and conidia.	Spots are small, spherical to elliptical with raised margins; black sporulation may be seen on the lesions.
Yellow halo around the spots	Usually lighter than those caused by *A. brassicicola*.	Usually darker than those caused by *A. brassicae*.	Translucent yellow halo.

(*Contd.*)

Table 16.1 The difference in *Alternaria* species (*A. brassicae*, *A. brassicicola*, and *A. raphani*). (*Contd.*)

Features	*Alternaria brassicae*	*A. brassicicola*	*A. raphani*
Host penetration	Penetrates leaf only through stomata.	Penetrates leaf directly or through stomata.	Direct penetration.
Conidia	Brownish black, obclavate, muriform, produced singly or in chains or 2–4.	Dark cylindrical to oblong, muriform, produced in longer chains of 8–20 or more.	Olive brown to dark, obclavate, muriform, more or less pin-pointed at each end, borne singly or in short chains of 2–3 spores.
Septa in conidia	Transverse septa: 6–19 Longitudinal or oblique septa: 1–8	Transverse septa: 1–10 Longitudinal septa: few	Transverse septa: 3–7 Longitudinal septa: several
Spore beak length (μ)	45–65	Beak absent	1–25
Host range	Wide	Wide	Narrow, common on radish

Disease Cycle and Epidemiology

Primary sources of pathogens are the infected seeds or non-decomposed plant debris in the top soil layers with over-wintering mycelium, spores (conidia and chlamydospores), and microsclerotial for years. However, *Alternaria* species are simple parasites that can survive outside the host, saprophytically. Seedborne inoculum can be externally on the seed surface or internally in seed tissue. Several *Brassica* weeds are also infected by the pathogens and can be another source of inoculum. During the vegetation period, the rain- and wind-transported conidia are also an important source of infection. Flea beetle (*Phyllotera cruciferae*) also transmits *Alternaria brassicicola* in cabbage. The conidia remain viable after passing through the digestive tract of the beetle. These fungi sporulate profusely and are disseminated during the growing season throughout fields by wind, splashing water, equipment, workers, and insects including flea beetles. Spread over long distances and new areas occur through infected seed. Most of the conidia from infected leaves of oilseed rape or other crucifer seed crops are released during harvesting and cleaning of crops; such released spores became airborne and carried over to vegetable plantings by the wind. The *Alternaria* blight of crucifers is favoured by warm temperatures (15°C–30°C), relative humidity of more than 90% for at least 12 hours, and wind velocity is 2–5 km/h and intermittent rains. At least 12 hours of 90% relative humidity are required for sporulation on plant tissues. Spore production is optimal at 18°C–24°C for *A. brassicae* and 20°C–30°C for *A. brassicicola*. Leaf wetness and free water is required for infection for at least 6–8 hours.

Management

Cultural Methods

- Use high-quality seeds free from the pathogen.
- Treat the seed if the seed is infected.
- Long rotations (3 years) without crucifer crops or cruciferous weeds such as wild mustard.

- Allow good air circulation (i.e. wide spacing, rows parallel to prevailing winds, not close to hedgerows).
- Destroy diseased crop residues after harvest to reduce inoculum sources.
- Do not plant vegetable brassicas close to field brassicas such as oilseed rape.
- For broccoli head rot, avoid using overhead sprinkler irrigation.
- Seed crops should be produced in arid climates and without the use of overhead sprinklers.

Physical Control

- Treat the seed with hot water at 50°C for 20 minutes.

Chemical Control

- Treat the seed with fungicides such as Thiram or Iprodione @ 2 g/kg of seed.
- Spray Zineb 75% WP @ 600–800 g in 300–400 litres of water/acre or Mancozeb 75% WP @ 600–800 g in 300 litres of water/acre.
- First foliar spraying with Tridemorph 0.1% followed by spraying with Mancozeb 0.25% at a month interval.
- Use effective foliar fungicides such as Iprodione, Triazoles such as Difenoconazole and Tebuconazole, and Strobilurin products.

16.2 BLACK ROT

Black rot is one of the most destructive diseases of vegetable Brassica crops worldwide. The disease is also known as black venation, bacterial wilt and tracheobacteriosis. Black rot is a systemic vascular disease. Black rot was first described by Harrison Garman (1894) as a disease of cabbage in Lexington Kentucky, USA. The pathogen was one of the first bacterium shown to be seedborne by Harding et al. (1904). In India, it was first reported by Patwardhan in 1928. Cabbage, cauliflower, and kale are the most susceptible hosts while broccoli, mustards, rape, rutabaga, and turnip are susceptible hosts. Radish, gobhi sarson (canola), and Chinese sarson are also infected. In warm and wet climates, the losses due to black rot may exceed 50% on some crops. Over 130 years later, black rot remains a threat to cabbage, cauliflower, and other *Brassica* crops around the world.

Symptoms

Typical disease symptoms include "V-shaped" chlorotic yellow lesions at leaf margins, blackening of the veins, which result from bacterial movement in the vascular system, necrotic tissues, premature leaf fall, stunted growth, and the death of young plants (Figure 16.2). Symptoms generally begin with chlorotic yellow lesions at the leaf margin, which later expand and produce the typical "V" shaped lesion, originating from entry points of thc bactcrium at hydathodes or wounds. Tissues within the chlorotic areas become necrotic (deadened) and leaf veins usually turn black, hence named "black rot". Systemically infected plants show stunted growth and develop more severe symptoms on one side of the plant. Seedlings produced from infected seeds become yellow, drop leaves prematurely, and may die soon after emergence, but serve as a source of inoculum for other healthy seedlings. The cross sections of systemically infected stems and leaf petioles exhibit a black ring due to blackening of vascular tissues, yellow slime droplets of bacteria, and sometimes, cavities filled with bacteria in the pith and cortex. Cabbage and cauliflower heads are invaded by bacterium through the systemic movement of vascular tissue, resulting in discoloured, unmarketable cabbage heads and prone to secondary infection during storage. The vascular discolouration and later internal break down is also shown by the infected fleshy tissue of radish, knoll-khol, turnip and rutabaga. Such fleshy infected tissues are often invaded by other soft rot causing bacteria and emit a repulsive foul smell.

Figure 16.2 Black Rot of Cauliflower. Typical V-shaped lesion on leaf blade.

Causal Organism and Etiology

Black rot of crucifers is caused by *Xanthomonas campestris* pv. *campestris* (*Xcc*). The bacterium is small (0.7–3.0 × 0.4–0.5μ), gram-negative, non-spore forming rods, with a single polar flagellum. It produces a yellow extracellular polysaccharide (EPS) called "xanthan" on glucose-containing media. While *Xcc* is an important plant pathogen, it is also used in industrial production of the polysaccharide xanthan, a thickening agent and stabiliser.

Disease Cycle and Epidemiology

Black rot is primarily a seedborne disease and bacterium (*Xcc*) can survive in seed for 3 years. A few, 3 infected seeds per 10,000 (0.03% infected seeds) can cause a black rot epidemic in the field. However, the bacterium can also survive in crop residues, infested soil, and cruciferous weeds species. The bacterium can survive in crop residue for upto 2 years. The bacterium can also infect cruciferous weeds which serve as reservoirs for the bacterium, these include pepper grass (*Lepidium virginicum*), black mustard (*Brassica nigra*), wild radish (*Raphanus raphanistrum*), wild turnip (*Brassica campestris*), wart cress (*Coronopus didymus*), and others. The bacterium commonly enters the plant through the hydathode or water pore, present on the leaf margin; however, leaf damage caused by insect feeding, hail, or mechanical injury and cultural operations can also enable pathogen entry. Bacteria multiply in loosely packed parenchyma cells and then enter xylem elements, where they move downward into the stalk and upward to leaf margins. Though hydathode infection is the most common, stomatal entry may occur when plants are subjected to heavy rains or irrigation. After the entry, the bacterium moves into the leaf veins and multiplies, rot and plug the veins. The black rot bacterium produces xanthan, an extracellular polysaccharide (EPS), which plugs the vascular tissue inside the veins that result in collapsing and blackening of the vein. The tissue above the plugged, collapsed xylem eventually turns yellow, wilts, and dies. The bacterial ooze or contaminated water droplets that exude through hydathodes or wounds from the infected leaf surface, disseminated by rain splashes, wind irrigation water, overhead irrigation, cultivation, insects or animals and tools, to other leaves. Secondary infection by other bacterial species can also contribute to the further development of severe rotting of vegetable tissue. Rain, heavy fog or dew, temperature 25°C–35°C, and high relative humidity (80%–100%) are very favourable for disease development. Symptomless infection can occur under cool and wet conditions.

Management

Cultural Methods

- Use of bacteria-free seed.
- Follow a three-year crop rotation with non-cruciferous crops.
- Transplant seedlings in soil free of black rot for at least two years.
- Remove and destroy crop debris and cruciferous weeds.
- Do not mow or clip transplants.
- Plant crops in well-drained soils and use irrigation practices that minimise leaf wetness.
- Keep fields free of cruciferous weeds.
- Disinfect seed beds and equipment with steam or germicidal sprays before use.
- Control insects to the minimise spread of the pathogen.

Physical Control

- Treat the seed with hot water (50°C for 30 minutes).
- Pre-drying of seeds at 40°C for 24 hours followed by air treatment at 75°C for 5–7 days has been found effective without causing seed damage.

Chemical Control

- Treat the seed with Tetracycline or Streptocycline or Agrimycin @ 0.1% helps to ensure a bacteria-free seed.
- Seed treatment with Aureomycin 1000 ppm for 30 minutes is effective in killing both the internally and externally seed-borne pathogen.
- Soak the seed in 0.5% Sodium hypochlorite (NaOCl) solution for 30 minutes.
- Drench the nursery soil with Formaldehyde @ 0.5%.
- Apply stable bleaching powder as soil drench @ 10–12 kg/ha to manage black rot and soft disease effectively.
- Sprays with Copper fungicides at 10-day intervals help reduce the spread of the disease.

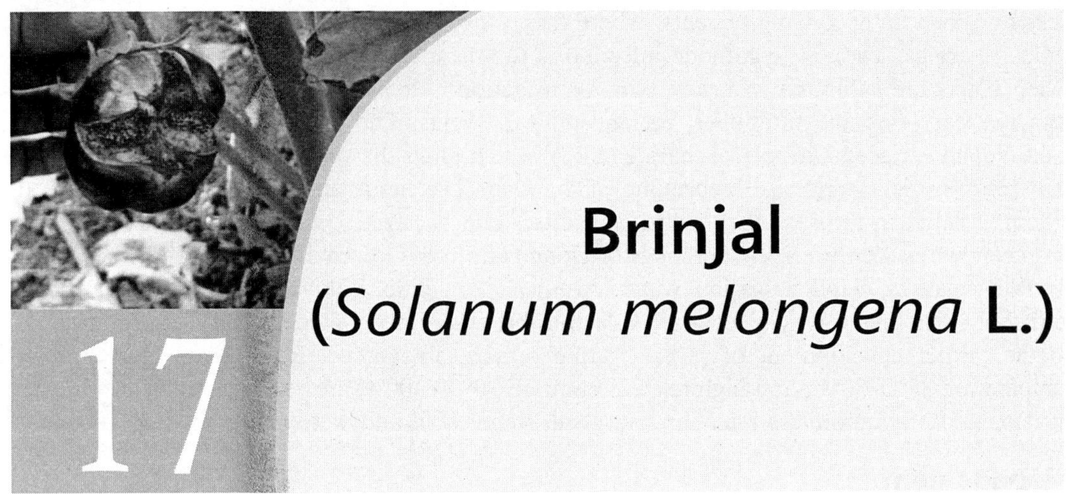

17 Brinjal (*Solanum melongena* L.)

17.1 PHOMOPSIS BLIGHT AND FRUIT ROT

Phomopsis blight and fruit rot is one of the major diseases of eggplant, especially in India. The disease was first reported in 1914 from Gujrat and since then from many parts of India. After fruiting, Phomopsis fruit rot causes yield losses of 15%–50%.

Symptoms

The symptoms of Phomopsis blight disease appear on leaves, stems, and fruit and it also can cause a seedling damping-off.

Seedling damping-off: Post-emergence damping-off of seedlings results from infection of the stem just above the soil surface.

Leaves: Initially lesions are small, more or less circular, and buff to olive, later becoming cinnamon buff, with an irregular blackish margin. Irregular spots result from coalescence. The lesions on the petiole or the lower part of the midrib can result in the death of the entire leaf. Later, numerous blackish, tiny pycnidia (conidiomata) develop in the center of older lesions. Affected leaves turn yellow and may drop prematurely.

Stem and branches: Elongated, blackish-brown lesions develop. If cracks and sunken cankers develop at the base stem, it can girdle the stem, and the shoot above the infected area wilts and dries up. Pycnidia develop readily in lesions on young stems, but rarely on older ones.

Fruit: Fruit lesions are sunken, discoloured, soft, and rot large areas of fruit. On these lesions, tiny black fruiting bodies (pycnidia) develop in concentric rings which are embedded in host tissues and cover most of the rotten fruit surface. If conditions become dry, infected fruit become dry, and form black mummies. If the infection enters the fruits, the whole fruit may shrivel and mummified due to dry rot. Such fruits are unmarketable.

Seeds: Discolourations of the seed occurs.

Whole plant: Seedling blight, damping off, dwarfing plant dead, dieback and topple over.

Causal Organism and Etiology

Phomopsis blight and fruit rot of brinjal is caused by *Phomopsis vexans*. It is a pycnidial fungus with sexual form in the genus *Diaporthe vexan* (ascomycetes). Mycelium is hyaline and septate. Pycnidia subepidermal, erumpent, dark, thick-walled, flattened to globose, with or without a beak. Conidiophores (phialids) are hyaline, simple or branched, occasionally septate. Pycnidium produces two types of conidia viz., alpha, and beta. Alpha conidia (or pycnidiospores) are hyaline, aseptate, subcylindrical. Beta conidia (or stylospores) are filiform, curved, hyaline, septate, non-germinating. Perithecia in culture usually form in clusters and beaked. Ascospores are biseriate, hyaline, narrowly ellipsoid to bluntly fusoid, one-septate (two-celled), and constricted at the septum.

Disease Cycle and Epidemiology

The fungus, *Phomopsis vexans,* survives in crop debris, seeds, and soil. The disease is primarily seedborne and seed infection occurs during fruit rot phase. *Phomopsis vexans* is a pycnidial fungus that produces the large number of conidia on conidiomata (pycnidia). Secondary spread of disease occurs by windborne alpha conidia which are dispersed by splashing rain, insects, and contaminated equipment. Spores germinate rapidly when free moisture is present on leaves or stems. The disease is favoured by warm, wet weather and is spread by splashing water. Disease is favoured by hot and wet weather. The optimum temperature for fungal growth ranges from 29°C–32°C and relative humidity for disease development is more than 55%.

Management

Cultural Methods

- Use pathogen-free seeds for planting.
- Transplants should be Phomopsis-free.
- Avoid continuous cultivation of brinjal.
- Summer deep ploughing helps in reducing initial inoculum.
- Adopt good field sanitation practices, destroy infected plant material, and follow crop rotation (3–4 years) with non-solanaceous crops such as paddy–sesame, to break the disease cycle.
- Mulching and furrow irrigation helps in reducing the splashing of water and soil.
- Remove and destroy the diseased crop debris.
- Keep the field weed free, since pathogens can survive on Solanaceous weeds, such as nightshades.

Physical Control

- Procuring seeds from healthy fruits followed by hot water treatment of seed at 50°C for 30 minutes.

Host Resistance

- Grow resistance/tolerant varieties such as JC-1, JC-2, Pant Samrat, Pant Brinjal Hybrid1, and Azad B-2.

Chemical Control

- Treat the seed with Thiram (2 g/kg seed) which protects the seedling in the nursery stage.
- Spray Zineb 75% WP @ 600–800 g in 300–400 litres of water/acre or Carbendazim 50% WP @ 120 g in 240 litres of water/acre.
- Spray Dithane Z-78 @ 0.2% or Bordeaux mixture @ 1%.

17.2 SCLEROTINIA BLIGHT

Diseases caused by *Sclerotinia* spp., generally known as 'white mold', can affect hundreds of plant species including many important crops. The disease of the crops caused by *Sclerotinia* spp. has been generally called Sclerotinia rot. The disease was also recorded as Sclerotinia stem rot, Sclerotinia fruit rot, fruit and stem rot, fruit rot, stem blight and rot, white mold or pink rot, and watery soft rot depending on the species of the solanaceous crops.

Symptoms

The symptoms appear on the stem, branches, leaves, and fruits. Primary symptoms generally appeared on main stem very close to the soil surface. The most characteristic symptom of the disease is white cottony mycelial growth of the pathogen on infected aerial tissues. The fungal hyphae produce enzymes and oxalic acid, creating water-soaked lesions. On opening dry portion of the stem, pith was full of fungal hard black sclerotia, elongated or cylindrical, medium to large in size and often attached end to end. Sclerotia are most commonly produced on the external surface of the infected tissue, but sometimes inside of soft host tissues or cavities such as floral receptacles, fruits and the pith of stems. Secondary symptoms such as wilting, bleaching, and shredding also can be observed on above-ground tissues including stems, leaves, petioles, and reproductive organs. In advanced stages, white, cottony mycelial mat covers the affected tissue, and sclerotia are produced on the fruit surface. Sclerotia are produced inside and outside of affected stems. Internal stem pith is fully filled with black sclerotia.

Causal Organism and Etiology

Sclerotinia blight of brinjal is caused by the fungus *Sclerotinia sclerotiorum*. The hypahe are closely septate, inter, and intracellular. Sclerotia are pinkish white when young, and become dark brown to black at maturity. Sclerotia vary in size and shape, being flattened, elongated or roughly elongated or roughly spherical, 2–12 mm in diameter. After germination sclerotia produce apothecia which are brownish or fawn coloured, funnel or cup shaped. In each apothecial cup, several asci develop. The asci are unicellular, cylindrical (108–152 × 4.5–10 μm) to ovate (7–16 × 3.6–10 μm). The ascospores are discharged violently through an apical pore in the ascus and are the chief means of the spread of the disease. The difference between two species of *Sclerotinia* (*S. sclerotiorum* and *S. minor*) is given in Table 17.1.

Table 17.1 Difference between *Sclerotinia sclerotiorum* and *Sclerotinia minor*

Features	*S. sclerotiorum*	*S. minor*
Sclerotia	Irregularly shaped, larger than a wheat grain (5–25 mm long).	Smaller and similar in size to raw sugar crystals (0.5–3 mm).
Mushroom-like fruiting bodies (apothecia)	Produces	Normally do not produce.
Dispersal	Spreads by sclerotia and by wind-blown ascospores.	Spreads by physical movement of the sclerotia.

Disease Cycle and Epidemiology

This fungus has a wide host range and survives as sclerotia in soil and in plant debris. Sclerotia buried in the plow layer of soil can survive and remain infective for up to 5 years. Dormant fungal resting structures called sclerotia on the soil surface or buried 0.75 to 1 inch germinate during cool, moist weather and form small mushroom-like structures called apothecia. These apothecia eject ascospores that are carried by wind and irrigation water to tomato and other hosts. The disease cycle begins primarily from windblown ascospores. Ascospores must have an energy source to initiate infection, and often this energy source is fallen flower petals. After the fungus grows on the dead or dying tissue it invades healthy stems. Several days of cool (12°C to 16°C) and continuously wet conditions are necessary for infection. Prebloom infection may occur if plants are damaged by frost or mechanically. The pathogen is readily disseminated by wind-blown spores, irrigation water, and in infested soil. *S. sclerotiorum* survives between tomato, pepper, and eggplant crops pathogenically on many other crops (dry bean, sunflower, melon, carrot, and potato among others), weeds, and as dormant sclerotia in the soil. Dew, fog, and frequent rain generally favour disease development. High soil moisture coupled with temperatures range of 16°C–21°C is favourable for disease development. In the absence of the host, the fungus subsists in the form of sclerotia in soil or affected plant debris.

Management

Cultural Methods

- Immediately after harvest, the affected plants and debris should be collected and burnt.
- Deep summer ploughing should be done in such a way that surface soil is buried deep.
- Rotation of cropping pattern with crops like beet root, onion, maize, paddy, and gingelly eliminate the fungal inoculum in the field.

Chemical Control

- Spraying Mancozeb 75% WP @ 3 g per litre or Carbandazim 50% WP @ 2 g per litre manage the disease.

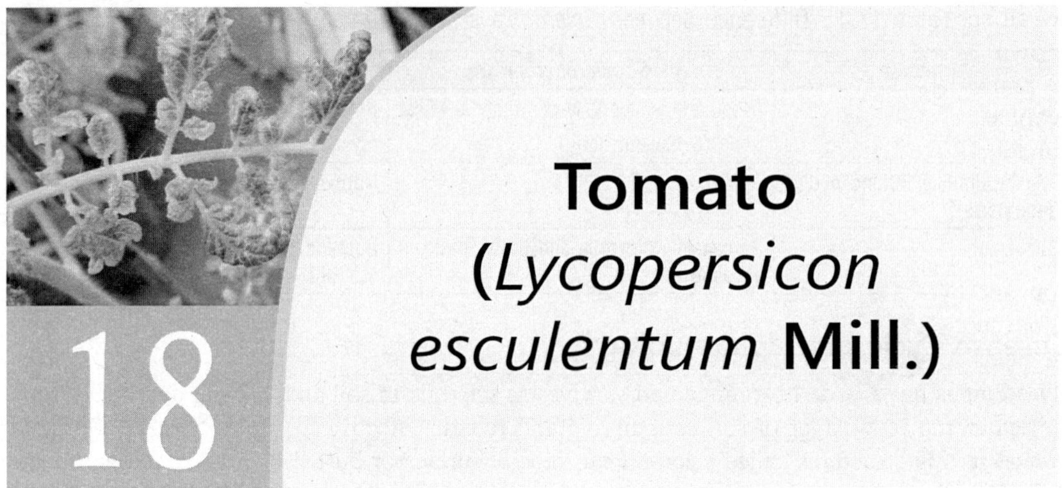

18 Tomato (*Lycopersicon esculentum* Mill.)

18.1 DAMPING OFF

Damping-off is one of the most important diseases of tomato causing high mortality of seedlings in the nursery and field.

Symptoms

"Damping-off" is a general term for the death of seedlings, either before (seedlings fail to emerge) or after emergence (seedlings fall over and die soon) after emergence, in cool and wet soils. These pathogens can also infect tomato fruit and cause field and postharvest fruit rots.

Pre-emergence damping-off: Seeds may rot before germination and seedlings may decay before emergence, giving the appearance of poor seed germination. The young radical and the plumule are killed and seedlings completely rot.

Post-emergence damping-off: The post-emergence phase is characterised by the infection of the young, juvenile tissues of stems that usually have tan to dark brown, water-soaked, soft lesions and shriveled portion at the soil line. With time the lower stem collapses, roots decay, cotyledons and leaves wilt and seedlings fall over and die. However, even if plants do not succumb, the surviving plant may be stunted and delayed in development.

Causal Organisms and Etiology

Tomato damping off is caused by the several pathogens such as *Phytophthora*, *Pythium*, and *Rhizoctonia*. *Phytophthora* and *Pythium* species are in the oomycete group. *Phytophthora capsici* and *P. parasitica* usually infect roots of older plants, but can cause damping-off. A number of *Pythium* species cause damping-off: *P. aphanidermatum*, *P. arrhenomanes*, *P. debaryanum*, *P. myriotylum* and *P. ultimum*. Sexual structures–antheridia, oogonia, and oospores–are produced by all species. *Rhizoctonia solani* has no asexual spores but produces characteristically coarse, brown, approximately right-angle branching hyphae. The hyphae are distinctly constricted at branch points, and cross walls with dolipore septa are deposited just after the branching. Hyphal cells are multi-nucleate. Small, tan to brown, loosely aggregated clumps of mycelia function as sclerotia.

Disease Cycle and Epidemiology

All these pathogens survive in the soil as saprophytes and infected crop residue. *Phytophthora* spp. and *Pythium* spp. produces chlamydospores and oospores for their survival while *Rhizoctonia solani* survives by forming thick hyphae and sclerotia. Most of these soilborne pathogens survive in fields for indefinite amounts of time. In addition to tomato, these pathogens can infect numerous other plants. *Phytophthora* spp. and *Pythium* spp. (except *P. ultimum*), usually produce zoospores that swim to and infect susceptible tissues. *Rhizoctonia solani* is a soilborne fungus with a very broad host range. This fungus can survive by infecting and thriving on a great number of plant hosts, besides tomato, and can also persist in the soil as a saprophyte. Wet soil conditions and cool temperatures (15°C–20°C) generally favour these organisms and their ability to grow and infect hosts. However, for pathogens such as *P. aphanidermatum* and *P. myriotylum*, warmer soil conditions (32°C–37°C) are favourable. Infection of *Phytophthora* and *Rhizoctonia* is also common under cool conditions, but these can also infect seedlings in warmer soils. The seedling stage is most susceptible to infection, though these pathogens can infect the feeder roots of mature plants. When tomato seedlings reach the 2- or 3-leaf stage, they are less susceptible to the infection by *Pythium* or *Rhizoctonia* damping off disease. However, *Phytophthora* damping off disease infects tomato plants at any stage.

Management

Damping-off is primarily managed by creating unfavourable conditions for the pathogens.

Cultural Methods

- Used raised seed bed.
- Provide light, but frequent irrigation for better drainage.
- Prepare soil to enhance drainage and avoid low areas.
- Stake plants or use plastic mulches on the bed tops to keep fruit off the ground.
- Avoid raising a nursery in a shad and humid place.
- Select well-draining nursery site away from tomato fields.
- Elevate seed beds to improve drainage.
- Transplant only disease-free seedlings.
- Avoid early planting of tomato seedlings as extensively infected crop residue still remains in the soil.
- Follow crop rotation as continuous tomato crops will increase the populations of the soilborne pathogens.
- Transplant seedlings in good quality beds and do not plant too deep.

- To avoid fruit rots, keep bed tops dry by carefully managing the irrigation or by using buried drip irrigation.
- Ensure good aeration through the wide spacing of seeds and staking of plants with wooden sticks.
- Avoid overwatering because this favours disease development.
- Uproot diseased seedlings. Burry deeply, or put into plastic bags and burn.
- Removal of top 30 to 60 cm of greenhouse soil and replacement with fresh uncultivated soil.
- Apply an optimum dose of nitrogenous fertilizers.
- Soil solarization and biofumigation are also useful.
- Disinfect seedling pots and trays with Sodium hypochlorite (NaOCl) solution before storage. Wear gloves or wash hands with water afterward, because bleach is toxic.

Biological Control

- Treat the seed with *Trichoderma viride* @ 10 g/kg of seed.

Chemical Control

- Seed treatment with fungicides or bioagents is the only preventive measure to manage the pre-emergence damping off.
- Treat the seed with Captan 75% WS @ 20–30 g/kg seed.
- Treat the seed with Thiram @ 3 g/kg of seed.
- Drench the soil with Copper oxychloride 0.2% or Bordeaux mixture 1% or Captan 75% WP @ 1 kg in 400 litres of water/acre in nursery.
- Spray Metalaxyl @ 0.2% in cloudy weather condition.

18.2 WILTS

The major diseases that produce wilting in tomatoes are Fusarium wilt, Verticillium wilt, and bacterial wilt.

18.2.1 Fusarium Wilt

Fusarium wilt is one of the most devastating diseases of tomato and has caused major losses in tomato production worldwide. In India generally, 30%–40% yield losses were recorded due to this wilt disease and sometime the loss may be gone upto 80% in favourable weather conditions for the disease.

Symptoms

The disease is characterised by vein clearing, drooping of leaves (epinasty), yellowing, wilting, browning, defoliation of the leaves, stunted growth, and eventually the death of plants. *Fusarium* often causes the yellowing of one side of plants or one side of leaves or branches. Generally, yellowing begins with the older, lower leaves, followed by wilting, browning, and defoliation. Plant growth is typically stunted, and little or no fruit develops. Infected plants often die before maturing. When an infected stem and large leaf petioles are split lengthwise browning of the vascular tissue is seen. The pith tissues of stems remain healthy. In wet weather, the pinkish fungal growth can be seen externally on dead plant tissues of affected stems and roots.

Causal Organism and Etiology

Fusarium wilt of tomato is caused by *Fusarium oxysporum* f. sp. *lycopersici.* Mycelium: septate, hyline; Macroconidia: fusiform, hyline, mostly 2–3 septate and less abundance than microconidia; no sporodochia, pinnotes, or sclerotia produced; microconidia: one-celled, hyaline, ovoid or ellipsoidal; terminal and intercalary chlamydospores produced on old mycelium.

Disease Cycle and Epidemiology

F. oxysporum f. sp. *lycopersici* survives in the soil for several years as a dormant propagule (chlamydospores) or saprophytically growing mycelium in infected plant debris or seed. Chlamydospores germinate in response to root exudates (e.g. amino acids) of the host. The infection hyphae adhere to the root surface and then penetrate directly or through wounds including those caused by the root-knot nematodes. On leaves, the pathogen penetrates through cuticles, stomata, or wounds. The mycelium invades the root cortical cells intercellularly and enters vascular system through the xylem pits. Subsequently, the fungus displays a unique pathway of infection where it tends to colonize exclusively inside the vessels of xylem; then rapidly colonize throughout the plant. The characteristic wilt symptoms appear due to the blockage xylem vessel by fungal growth (mycelium), abundant sporulation, toxins (e.g. Fusaric acid), gums, gels, and formation of tyloses. Secondary spread of the pathogen from one plant to another seldom occurs. Dissemination of the pathogen can occur via seeds, infected transplants, soil, or other means. Disease development is favoured by warm soil temperatures ranging from 27°C–32°C, dry weather, sandy soils, acidic soil (pH 5–5.6), and plants preconditioned with low nitrogen and phosphorus, and high potassium, both in greenhouse and field conditions. The optimum temperature for disease development is 28°C. In soil infested with wilt fungus and root knot nematodes, wilt occurs at significant level at the temperature range 16°C–24°C. Virulence of the pathogen is enhanced by micronutrients, phosphorus, and ammonium nitrogen and decreased by nitrate nitrogen.

Management

Cultural Methods

- Use pathogen-free seeds for planting.
- Remove the affected plants.
- Avoid over-application of high nitrogen fertilizers. High soil nitrogen levels accompanied by low potassium levels can increase susceptibility to the fungus.
- Avoid cultural activities when soil and plants are wet.
- Sanitize stakes and tomato cages at the end of the season. A thorough cleaning with water will reduce most risk of transmitting the disease.

- Follow long (5–6 years) crop rotation with a non-host crop such as cereals, onion, garlic, tobacco, okra, spinach and cauliflower. Cowpea and Sesbania are symptomless carriers. Do not rotate the tomato crop for upto 4 years with other Solanaceous crops such as potato, pepper, and eggplant as these all are susceptible to the wilt disease.
- Adjust the soil pH to 6.5–7.0 and using Calcium nitrate as fertilizer.

Host Resistance

- Use resistant/tolerant varieties for planting.
- Due to the presence of races of the pathogen, often resistant varieties fail. The term "race" refers to the organism's ability to infect certain crop varieties but not others. The three known races (Races *Fol* 1, 2, and 3) of *F. oxysporum* f. sp. *lycopersici* on tomato cultivars are distinguishable by their principle resistance genes.

Biological Control

- Seed treatment with *Trichoderma viride* 1% WP @ 9 g/kg seed
- Root zone application: Mix thoroughly 2.5 kg of the *T. viride* 1% WP in 150 kg of compost or farmyard manure (FYM) and apply this mixture in the field after sowing or transplanting of crops.

Chemical Control

- Treat the seed with Carbendazim @ 2.5 g/kg of seed against seedborne inoculum.
- Spot drench with Carbendazim (0.1%).

18.2.2 Verticillium Wilt

Verticillium wilt is a devastating disease affecting many economically important crops including tomato. A 100% incidence of Verticillium wilt has been observed on tomato cultivar "Bonny Best" grown in soil artificially infested with at least 200 microsclerotia of *V. dahliae* per gram of soil.

Symptoms

The typical symptoms of Verticillium wilt start on the lower leaves, with chlorosis and characteristic "V-shaped" necrotic chocolate brown lesions in between the veins or at leaf margins with yellow halos that expand to cause browning or purpling of veins. The leaves may wilt, die, and drop off. Frequently, leaves on all sides of the plant uniformly show symptoms. As the disease progress, plants show stunted growth, wilting of branches or the entire plant, and eventually die. Verticillium wilt also causes discolouration (dark brown or black) of the vascular system, mainly in the lower stem, almost identical to that in Fusarium wilt except that the browning does not extend up the stem more than 10–12 inches above the soil. Plants wilted during the day (hot weather) may recover at night. Discolouration does not occur in the petioles. However, the discolouration can be traced into the roots, upward and downward. Small, black, thick-walled, resting structures, called microsclerotia, may be present in the dying tissue. Like Fusarium wilt, this disease appears first on the lower leaves and progresses upward. Yellow blotches develop on lower leaves; the leaves rapidly turn completely yellow, wither, and drop off. Unlike Fusarium wilt, symptoms of Verticillium wilt do not progress along one side of a leaflet, branch, or plant.

Causal Organism and Etiology

Verticillium wilt of tomato is caused by *Verticillium dahlia* and *Verticillium albo-atrum. Verticillium* belongs to the fungal class Deuteromycetes (Fungi Imperfecti), a group of fungi that do not have a known sexual stage. The vegetative mycelium is hyaline, septate, and multinucleate. The nuclei are haploid. Conidia are ovoid or ellipsoid and usually single-celled. They are borne on phialides, which are specialized hyphae produced in a whorl around each conidiophore. Each phialide carries a mass of conidia. Verticillium is named for this "verticillate" (=whorled) arrangement of the phialides on the conidiophore. As the diseased plant senesces, the fungus becomes saprophytic and colonizes the dying tissues. During colonization, the fungus forms microsclerotia, which are masses of melanized hyphae.

Disease Cycle and Epidemiology

Verticillium is a notorious soil-borne pathogen. The disease cycle of both species is similar in most aspects except that *V. dahliae* produces microsclerotia and *V. albo-atrum* produces melanized mycelia for survival between crops. *Verticillium dahliae* is a hemibiotrophic fungal pathogen. The disease cycle starts with microsclerotia, a resting structure in soil or crop debris that is capable of surviving without a plant host for more than a decade. The microsclerotia germinate in the presence of root exudates of both host and non-host plants. Infection is enhanced if rootlets are broken or nematodes feeding on the root systems. The fungal mycelium infects roots through root tips or sites of lateral root formation and colonizes the cortex. From the cortex, the hyphae invade the vascular system where it grows rapidly and sporulates in the xylem vessels or sap-conducting channels. This results in interference with the normal upward movement of water and nutrients. The fungus produces a toxin (VD toxin) that contributes to the wilting and spotting of the leaves. Verticillium wilt is favoured by moist soils and a temperature range of 21°C–27°C.

Management

Cultural Methods

- Tomato seedlings should be obtained from Verticillium wilt-free nurseries.
- Avoid the site for planting tomato with Verticillium wilt history.
- Follow 3–4 years crop rotation. Cereal crops, e.g. corn and grains are non-hosts and are recommended in the rotation.
- Remove weeds on regular basis.
- Do not plough under the debris as the fungus can survive in the soil.
- Manage the nematode *Pratylenchus penetrans*, if present.
- Apply nitrogen and phosphorus at optimal rates to reduce the severity of Verticillium wilt.
- Avoid overwatering, especially early in the growing season.

- Remove and destroy crop residue after harvesting.
- Deep ploughing and even complete soil inversion can be effective in reducing disease losses as propagules of *V. dahliae* are generally present in the top 30 cm of soil.
- Use sanitized implements and equipment to prepare the soil for planting.

Physical Control

- Soil solarization is the procedure aimed at increasing soil temperature by means of round cover with polyethylene or PVC sheets to trap solar radiation into the soil has been proven to be effective against *V. dahliae*, other soilborne pathogens, nematodes and weed seeds.

Host Resistance

- In the presence of very high soil populations of *Verticillium*, however, even resistant cultivars may exhibit symptoms of Verticillium wilt.

Biological Control

- Seed treatment with *Trichoderma viride* 1% WP @ 9 g/kg seed
- Root zone application: Mix thoroughly 2.5 kg of the *T. viride* 1% WP in 150 kg of compost or farmyard manure and apply this mixture in the field after sowing or transplanting of crops.

Chemical Control

- Systemic fungicides, particularly benzimidazoles (i.e. Carbendazim, Benomyl, Thiabendazole, and Thiophanate-methyl) have been found effective against Verticillium wilt of tomato.
- Treat the seed with Carbendazim @ 2.5 g/kg of seed against soilborne inoculum.
- Spot drench with Carbendazim (0.1%).

18.2.3 Bacterial Wilt

Bacterial wilt is one of the major diseases of tomato and other Solanaceous plants and causes huge economic loses worldwide. The disease is known to occur in the wet tropics, sub-tropics, and some temperate regions of the world. This disease causes very heavy losses varying from 2% to 90% in different climates and seasons in India. The disease is also known as Southern bacterial wilt. In India, the disease was first reported by Cappel in 1892 from Pune district of Maharashtra as "Bangle blight" or "Bungdi disease" of potato; although its bacterial nature was established in 1909 by Coleman. Hutchinson in 1913 reported it as Rangpur wilt of tobacco. It is the first bacterial disease recorded in India.

Symptoms

Tomato bacterial wilt is commonly observed as wilting of one or half of a leaflet, at any growth stage of plants, during the hot and humid conditions, and plants may appear to recover at night. Initial symptoms of bacterial wilt are a loss of turgidity, drooping of the youngest leaves at the ends of the branches and stems, giving the plants a flaccid appearance (Figure 18.1). As the disease progress, the entire plant may wilt suddenly without yellowing and desiccating, although the dried leaves remain green, leading to permanent wilting and yellowing of foliage and eventually plant death. Another common symptom is the stunting of plants. Affected roots decay, becoming dark brown in colour and if the soil is moist, diseased roots become soft and slimy. Brown, sunken cankers are often visible at the base of the plant. In young tomato stems, when an infected stem is cut across infected vascular bundles may become visible as long, narrow, dark brown streaks. The discolouration of vascular tissues in the roots and main stem extends upto 10 inches above the ground. A common sign of bacterial wilt of tomato can be observed, at the surface of freshly-cut sections from severely infected stems, as a slimy, sticky, milky-white to brownish beads (ooze), which indicates the presence of dense masses of bacterial cells in infected vascular bundles, and particularly in the xylem. Difference in Fusarium wilt, Verticillium wilt and bacterial wilt of tomato is given in Table 18.1.

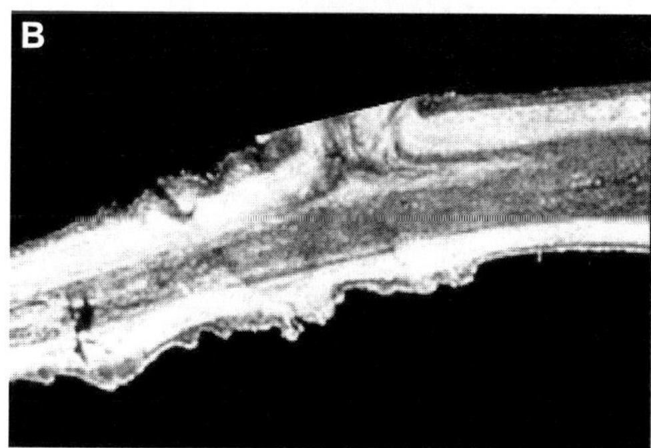

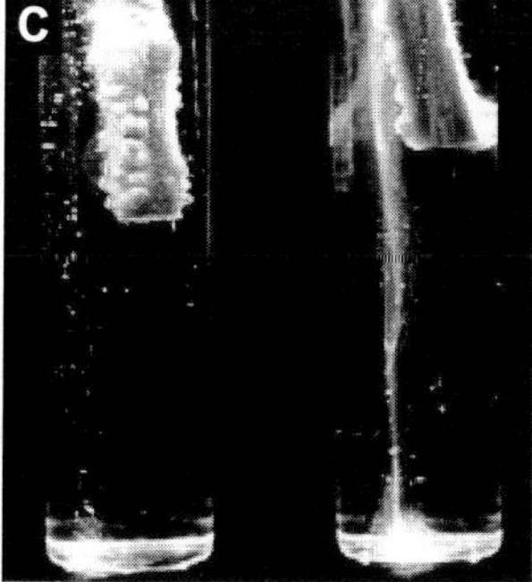

Figure 18.1 Bacterial Wilt of Tomato. (A) Drooping of leaves (*Courtesy:* Clemson University-USDA Cooperative Extension Slide Series, Bugwood.org), (B) Brown discoloration of the vascular system of stem, (C) Classical Ooze Test: Bacteria streams or ooze flows from a cut of infected stem (right), but not from a healthy stem (left). (*Courtesy:* Ray Cerkauskas, AVRDC, 2004).

Table 18.1 Difference in Fusarium wilt, Verticillium wilt and bacterial wilt of tomato

Features	Fusarium wilt	Verticillium wilt	Bacterial wilt
Causal organism	*Fusarium oxysporum* f. sp. *lycopersici*	*Verticillium dahliae, Verticillium albo-atrum*	*Ralstonia solanacearum*
Symptoms	Fusarium wilt begins with lower leaves wilting or drooping, turning yellow then brown, and dying. Often just one branch or branches on one side of the plant will show symptoms. Leaflets on one side of the petiole may be affected.	The leaves may turn yellow, starting at the edges and between the veins, and spreading inward. Usually, both sides of the leaf are affected. Leaflets may also show a characteristic V-shaped or fan-shaped yellowing, with the widest part of the V on the leaf edge.	Sudden wilting without yellowing and leaves remain green.
Pith symptoms	Tomato seedling with a stem cut shows the browning of the vascular system and pith.	When diseased plant stems are cut open with a knife and exposed, the water-conducting tissues usually appear streaked with a dark brown or black discolouration. Discolouration does not occur in the petioles in verticillium wilt.	This resembles the symptoms of bacterial wilt, but a bacterial ooze test will be negative for Fusarium wilt.
Pattens	Symptoms first noticed on the lower (older) leaves, yellowing and wilting on one side of the plant or leaf.	Symptoms first noticed on the lower (older) leaves; uniform yellowing and wilting of the lower leaves.	Symptoms first noticed on the upper (younger) leaves (top-down wilt).
Virulence factor	Toxins (Fusaric acids)	VD-toxin	Extracellular polysaccharides (EPS) and endoglucanase.
Resting structures	Chlamydospores, sclerotia absent.	Small, black, thick walled resting structures, called microsclerotia, are formed in the dying tissue.	—
Growth stage at which plants affected	At any growth stage.	Near maturity	At any stage of plant growth.
Favorable conditions	More common during warm weather (soil temperature 27°C–32°C).	More often when temperatures are cool (21°C–27°C).	High moisture and high temperature (30°C–35°C).

Causal Organism and Etiology

Tomato bacterial wilt is caused by *Ralstonia solanacearum*. It is a Gram-negative, aerobic, rod-shaped bacterium measuring 0.5–0.7 × 1.5–2.0 μm in size and motile by polar tuft flagella. It is a destructive soil-borne pathogen in many economically important crops like tomato, eggplant, groundnut, pepper, and potato. Strains of *R. solanacearum* have conventionally been classified as races and biovars. Bacterial wilt of tomato is caused by either race 1 or race 3 of *R. solanacearum* and, rarely, by race 2. Race 1 is endemic in the United States.

Host range: *Ralstonia solanacearum* causes bacterial wilt and infects over 200 plant species in 50 families.

Disease Cycle and Epidemiology

Ralstonia solanacearum can survive in diseased plants or plant debris, wild hosts, and seeds. The pathogen can also thrive for up to 40 years in pure water. *R. solanacearum* infects host plants primarily through wounds formed by lateral roots emergence or by damage caused by nematode feedings. The pathogen can also enter plants from stem injuries caused by insects or mechanical damage. Once inside the host, the bacterium has an affinity to the vascular systems, where it multiplies, filling the xylem elements rapidly with bacterial cells and slime. Xylem-dwelling *R. solanacearum* releases a virulence factor, extracellular polysaccharides (EPS), and cell wall degrading polygalactouronases (endoglucanase) that increase the viscosity of the xylem fluid, block the xylem vessels, and hinder the water transport; making the plant succumb to bacterial wilt. Wilting occurs 2–5 days after infection depending on host susceptibility, temperature, and virulence of the pathogen. Infection and disease development are favoured by high temperatures (optimum 30°C–35°C), high moisture and all types of soil including clay, and sandy with slightly acidic pH (pH less than 7.0). Race 3 biovar 2 is most severe on plants between 24°C and 35°C and decreases in aggressiveness when temperature exceeds 35°C or falls below 16°C.

Management

Cultural Methods

- Intercropping or incorporation of green manure crops such as hemp (*Crotovora juncea* L.) and mungbean (*Vigna radiata* L.) have been reported to be effective against bacterial wilt.
- Avoid growing tomato crops in the same field season after season.
- Avoid the plot infested by bacterial wilt.
- Remove and destroy infected plants during and at the end of the cropping season.
- Avoid planting tomato after tomato, brinjal, chili or potatoes.
- Rotate crop regularly or for atleast 5–7 years using non–susceptible crops.
- Plant tomato in raised beds.
- Space plants out evenly to improve air circulation
- Test soil, if the soil has a high pH, then adjust the pH to 6.2–6.5, using lime (Calcium carbonate) @ one tonne/ha, which is suitable for growth of tomatoes and most garden vegetables.
- Wash hands and gardening tools after handling infected plants.
- Rotate the tomato crop with non-host crops such as rice or green manures, beans, corn and cabbage.
- Flood the field 1–3 weeks before planting tomato.
- Provide additional space between the plants for air-circulation freely.
- Use pathogen-free seedbeds for raising clean transplants. This can be achieved by fumigating or treating seedbeds.
- Reduce root knot nematodes in the field which aggravate bacterial wilt problems.

- Plant seedlings in raised beds to improve drainage around the roots.
- Prevent movement contaminated water from infected fields to other the field.
- Drain the field quickly after the rain.
- Disinfect pruning tools either by (i) soaking them for at least 5 minutes in any disinfectant solution such as household bleach, pine oil cleaner, lysol or dettol and rinse them thoroughly with clean water and allowing them to dry before use or wash with water or (ii) bleaching or sterilizing them with a flame.
- Silicon induces basal resistance in tomatoes against *Ralstonia solanacearum* by modifying pectic cell wall polysaccharide structure.
- Calcium nutrition of tomato affects resistance to wilt and decreases the disease severity in a moderately resistant cultivar.

Host Resistance

- Grow bacterial wilt resistant/tolerant tomato verieties such as Arka Ananya, Arka Abhijit, Arka Abha, Arka Alok, etc.
- Use of resistant tomato or eggplant as rootstock for grafting can increase the resistance level of the scions. Grafting can be very effective against a variety of soil-borne viral, bacterial, fungal and nematode diseases.

Biological Control

- Dipping the seedling roots in a biopesticides solution mix of *Trichoderma harzianum* + *Pseudomonas flourescens* @ 10 g/L of water will protect the seedlings initially from seedborne fungal and bacterial diseases such as damping off, root rot, crown rot, and bacterial wilt.
- Apply neem cake @ 250 kg/ha one to two week before transplanting.

Chemical Control

- If infection occurs after transplanting, remove the infected plants and drench the soil with copper oxychloride solution @ 2 g/L of water.
- Spray the crop with Streptocycline @ 0.3 g/L of water, 2–3 times at 10 days intervals after transplanting.
- Apply plant resistance inducer, such as Acibenzolar-S-methyl (Actigard 50 WG) which enhances resistance against bacterial wilt disease if it is used in combination with moderately resistant cultivars.
- Chlorinate irrigation water continuously if using surface water or *R. solanacearum* infested pond water.
- Dip seedlings in the solution of fungicide Thiram or Carbendazim @ 2.5 g/L followed by dipping in Thiamethoxam 25 WG @ 1 g/L water for 10–15 minutes.

18.3 EARLY BLIGHT

Early blight is one of the dreadful diseases and major economic constraints to tomato production worldwide, especially in subtropical and tropical regions. Early blight is the most destructive disease which affects fruits and foliage resulting in a loss of about 80% yields.

Symptoms

Early blight symptoms appear on all parts of the plant including the leaf (leaf blight), stem (collar rot or stem lesions), and fruit (fruit rot) of tomato.

Leaves: Early blight symptoms typically appear first as small, circular to angular, dark brown to black, necrotic spots surrounded by yellow margins, on the older leaves. As these lesions enlarge, the characteristic concentric rings develop giving lesions a target board-like appearance. Several leaf spots coalesce; the entire leaflets dry up and collapse (Figure 18.2). Heavy defoliation increase the rate of respiration while decrease the photosynthetic rate. Defoliation also exposes fruits to the sun, causing sunscald. This leads to poor fruit quality and significant yield loss. Older leaves are more susceptible than younger leaves.

Figure 18.2 Early Blight of Tomato. (A) Brown spots on tomato leaves, (B) Concentric rings with yellow halo on leaf.

Stems: On the stems, elongated, oval to irregular, dry, sunken, dark brown to black lesions with concentric rings appear. If the infection girdles the stem, the seedling wilts, and dies. In young seedlings, lesions at the ground level girdle the stem; destroy the vascular system and cause "collar rot". On the fruits, the spots are dark brown or black, sunken, extending over part or all of the fruit. The rots become covered in black spore masses of the fungus.

Fruit: Infection on green or semi ripe fruits, causes dark, sunken decay, leathery lesions with distinct concentric rings, and usually covered with a black mass of spores, near the fruit stalk. These lesions enlarge and extend deep into the flesh of the fruit. Infected fruits mostly drop prematurely.

Causal Organism and Etiology

Early blight of tomato is caused by several species of *Alternaria* including *Alternaria linariae* (which includes *A. solani* and *A. tomatophila*), as well as *A. alternata*. *Alternaria* produces unique club-shaped conidia, often beaked with horizontal and often vertical septa that may be produced either individually or in a chain, depending on the species. Hyphal cells are darkly pigmented with melanin, which guards hyphae and spores against environmental stress and allows spores to survive in soil for long periods of time. Pathogen produces toxins (Solanapyrone A and Alternaric acid).

Disease Cycle and Epidemiology

The fungus overwinters in soil, plant debris, seed, and alternate hosts in the form of either conidia or thick-walled chlamydospores or dark pigmentation mycelia, which may serve as primary sources of inoculum. Dark pigmentation of the hyphae provides resistance to lysis. Conidia germinate at a temperature of 8°C–32°C in cool and humid conditions in the presence of free moisture to form germ tubes. Germ-tubes grow to form appressoria and penetrate the epidermis directly or indirectly through wounds or stomata thereby causing infection. Lesions appear after 2–3 days of infection depending on environmental conditions, leaf age, and cultivar susceptibility, and spores are produced 3–5 days after the appearance of lesions. Generally, a long period of wetness is needed for sporulations, but spores are also produced during alternate wet and dry conditions. Secondary spread of inoculum (conidia) occurs by wind, windblown soil, splashing rain, overhead irrigation and irrigation water but occasionally, flea beetles, and continues the disease cycle in other healthy parts of the same plant or different plants. Early blight is considered polycyclic with its repeating cycles of new infection and short disease cycle. The primary symptoms appear as necrotic with chlorotic, yellow halos. Mycelia from necrotic lesions that produce conidia that invade healthy leaves begin to produce secondary infections. As the conidia are unable to infect directly or by the formation of appressoria it invades through wounds and stomata and infects tuber and periderm. Infection reduces significantly by wound healing, by suberisation and by the formation of abscission layers. Warm, wet weather favours the rapid spread of early blight. Infection occurs during warm and humid conditions. Growth of the germ-tube requires a long wetting period than that of spore germination. Germ-tubes grow to form appressoria and invade the epidermis directly or indirectly through wounds or stomata. Optimum inoculum dose and wetting hours, the minimum temperature of less than 10°C, the maximum more than 35°C and the optimum range between 20°C–30°C is necessary for infection. The disease is favoured by dews and rainfalls, and is most severe on weak plants resulting from poor soil fertility, drought, or other pests.

Management

Cultural Methods

- Use pathogen free-seeds for planting.
- Collect seeds only from disease-free plants.
- Crop rotation with non-host crops for at least two years.
- Remove and burn infected plant residues.

- Use healthy seedlings for transplanting.
- Change the location of nursery beds in every season.
- Eradicate weed hosts and volunteer tomato plants.
- Apply a recommended dose of fertilizers. Maintain the fertility at optimal levels-nitrogen and phosphorus deficiency can increase susceptibility to early blight. Spray urea @ 1% at 45 days and another after 10 days.
- Irrigate early in the day to promote rapid drying of foliage.
- Avoid working in fields when plants are wet from rain, irrigation, or dew.
- Use drip irrigation instead of overhead irrigation to keep foliage dry.
- Stake the plants to increase air circulation around the plants that facilitate drying. Staking will also reduce contact between the leaves and spore-contaminated soil.
- In greenhouse production, early blight has been reduced upto 50% by covering houses with UV-absorbing vinyl film.
- Apply plastic or organic mulch to reduce humidity and provide a barrier between contaminated soil and leaves.
- Remove or bury infected plants to reduce the likelihood of the pathogen surviving in the following year.

Biological Control

- Extracts of several plants such as *Cinnamomum zeylanicum*, *Syzygium aromaticum*, *Ferula foetida*, *Inula racemosa*, *Hemidesmus indicus*, *Rubia cordifolia*, *Glycyrrhiza glabra*, and *Saussurea lappa* has been shown antifungal activity against *Alternaria solani*.
- In addition, the role of fungi such as *Trichoderma viride* and *T. harzianum* in the management of early blight has been reported.

Host Resistance

- Grow resistant/tolerant varieties such as Arka Samrat, Arka Rakshak, Arka Abhed, Arka Apeksha (H-385), Arka Vishesh (H-391), and Arka Vardan.

Chemical Control

- **Seed treatment:** Treat seed with Thiram @ 2.5 g/kg seed before sowing.
- **Foliar spray:** Spray any one of the following fungicides as per recommended dose given below:

Fungicides	Recommended Rate
Copper oxychloride 50% WP	1000 g in 300–400 litres of water per acre
Azoxystrobin 23% SC	200 ml in 200 litres of water per acre
Captan 50% WP	1000 g in 300–400 litres of water per acre
Captan 75% WP	666.8 g in 400 litres of water per acre
Iprodione 50% WP	600 g in 200 litres of water per acre
Kitazin 48% EC	80 ml in 80 litres of water per acre

(*Contd.*)

Fungicides	Recommended Rate
Mancozeb 35% SC	200 ml in 200 litres water per acre
Mancozeb 75% WG	400 g in 200 litres of water per acre
Pyraclostrobin 20% WG	150–200 g in 200 litres of water per acre
Zineb 75% WP	600–800 g in 300–400 litres of water per acre
Ziram 80% WP	600–800 g in 300–400 litres of water per acre
Famoxadone 16.6% + Cymoxanil 22.1% SC	200 ml in 200 litres of water per acre
Metiram 55% + Pyraclostrobin 5% WG	600–700 g in 200 litres of water per acre
Metriam 70% WG	1000 g in 200–300 litres of water per acre

18.4 LATE BLIGHT

Late blight has been identified as a major disease of tomatoes and potatoes and is one of the most devastating plant diseases of all time. Late blight is the disease that caused the Irish potato famine of the 1840s. The pathogen was first described by M.J. Berkeley as *Botrytis infestans* and subsequently named *Phytophthora infestans* by Anton de Bary (the "father of plant pathology") in the 1870s. This scientific breakthrough gave rise to the discipline of "Plant Pathology". An unprotected tomato field can suffer yield losses reaching up to 100% because of late blight infection. Late blight is a potentially serious fungal disease of potato and tomato. The fungus-like microorganism, *Phytophthora infestans*, was responsible for the infamous Irish potato famine of the 1840's. In 1861, Anton deBary proved conclusively that *Phytophthora infestans* actually caused the disease.

Symptoms

Late blight symptoms appear on all foliar parts including the leaves, petioles, stems and fruits.

Leaves: On the leaves, symptoms appear as pale green, irregular water-soaked spots, often begins at tip or edge of the younger, more succulent leaves (Figure 18.3A). These lesions are often surrounded by a pale yellowish-green border or halo. Under favourable conditions, lesions enlarge rapidly and turn dark brown to purplish-black, shrivel and killing the leaves instantly. Lesions are visible on both sides of the leaves. High humidity and leaf wetness, favour the growth of a cottony, white mold growth containing mycelium, sporangiophores, and sporangia, on the lower side at the edges of lesions. However, in dry weather, infected leaf tissues quickly dry and the white mold growth disappears. Infections progress through leaflets and petioles, resulting in large sections of dry brown foliage.

Stems and petioles: Lesions begin as indefinite, water-soaked spots that enlarge rapidly into brown to black lesions that cover large areas of the petioles and stems. During wet weather, lesions may be covered with a grey to white moldy growth of the pathogen. Affected stems and petioles may eventually collapse at the point of infection, leading to the death of all distal parts of the plant.

Figure 18.3 Late Blight of Tomato. (A) Leaf symptoms, (B) Fruit symptoms, (C) Severely affected tomato field.

Fruits: The pathogen produces grey-green, olivaceous greasy, water-soaked spots on green fruits which enlarge, coalesce, and darken, resulting in large, firm, chocolate brown, leathery lesions covering the entire fruit, rendering them unmarketable (Figure 18.3B). Under wet and high humid conditions, a thin layer of white mycelial growth may develop on fruit lesions. Infected tomato fruit may be invaded by secondary pathogens, causing soft-rot and disintegration of fruit. Difference in early blight and late blight is given in Table 18.2.

Table 18.2 Difference in early blight and late blight

Features	Early Blight	Late Blight
Causal organism	*Alternaria* spp. (*A. solani*, *A. alternate*, and *A. tomatophila*)	*Phytophthora infestans*
Symptoms	Lesions tend to be dark brown in colour with concentric rings.	Lesions caused by late blight tend to be light brown or tan in colour.
Presence of Lesions	Typically starts on the lower (older) leaves and slowly moves up the plant.	Anywhere on plants, but are mostly found on the young leaves (new growth).
Life style	Necrotroph	Hemibiotroph
Disease cycle	Polycyclic	Polycyclic
Pathogen	True fungi	Fungi like organism
Favorable conditions	Optimum temperature is 25°C–30°C.	High moisture, cool temperature (10°C–24°C)

Causal Organism and Etiology

Phytophthora infestans cause late blight of tomato and potato. *Phytophthora infestans*–literally, "plant destroyer", in Greek–has been traced back to the same origin as tomatoes and potatoes. *P. infestans* is a coenocytic oomycete with rare cross walls. Asexual reproduction is via sporangia that are ellipsoid to lemon-shaped with a small pedicel. Sporangia are 29–36 × 19–22 μm. Sporangia germinate either directly to form a germ tube (at temperatures of 15°C–25°C), or indirectly via zoospores (at temperatures below 15°C). Zoospores (7–12 per sporangium) have two flagella, one forward-directed tinsel type and a backward-directed whiplash type (heterokont). Zoospores are usually uninucleate, but binucleate zoospores have been detected.

Disease Cycle and Epidemiology

The fungal-like organism that causes late blight, *Phytophthora infestans*, affects both tomatoes and potatoes. *Phytophthora infestans* survives in infected crop debris as a mycelium and oospores in soil and potato tubers. The asexual disease cycle consist of: sporangium → germination → zoospore → encystment → germination → appressorium → host penetration and development of infection vesicle → intercellular hyphal growth → haustorium formation → initiation of sporulation → dispersal and can occur in fewer than five days. Multinucleate sporangia and uninucleate motile zoospores (water-swimming spores) represent the primary dispersal stages which are dispersed through wind, rain, or wind-blown rain. The disease cycle begins with the landing of sporangia on host plant tissue, a film of water must be present until the establishment of infection; otherwise, infection will not occur. To initiate infection, the germination of sporangia occurs either through the direct extension of germ tubes or by zoosporogenesis. Direct germination of sporangium on host tissue occurs by producing germ tube at temperatures above 21°C (optimally at 25°C). Sporangia germinate indirectly by producing up to eight biflagellate zoospores at cool, moist conditions and temperatures below 15°C, with an optimal temperature 12°C. Zoospores move through the water film, detach their flagella, and encyst until producing germ tubes. Germ tubes differentiate into appressoria that penetrate the host through the leaf cuticle, or less frequently, the stomata.

The life cycle of *P. infestans* stages involved sexual and asexual reproduction, propagule dispersal, spore germination, host penetration, and biotrophic or necrotrophic phases of infection. *P. infestans* adopts two different life-styles, typical of hemibiotrophs. During the early biotrophic growth phase, nutrients are obtained from living plant cells. This is achieved by forming a penetration peg, which pierces the cuticle and enters an epidermal cell to form an infection vesicle. Formation of digitate haustoria (i.e. appressoria-like structures) by the branching hyphae, and the subsequent secretion of enzymes that degrade components of the plant cell wall further facilitate the nutrient access. This is followed by extensive necrosis of host tissue resulting in colonization and sporulation. During necrotrophic growth, nutrients required for pathogen growth and reproduction are obtained from dead and dying cells in the necrotic lesions that develop as the pathogen colonizes the plant. As the infected tissue necrotizes, the mycelium develops sporangiophores that emerge through the stomata to produce numerous asexual sporangia. Each late blight lesion can produce 100,000–300,000 sporangia (2N) per day, contributing to a rapid spread of the disease. After sporulation, eventually, zoospores release from sporangia to promote aerial transmission of the pathogen and continue the disease cycle. *P. infestans* reproduces very rapidly and can destroy an unprotected tomato crop within a few days of occurrence. Dispersal of the pathogen is by wind, rain, or human-assisted via movement of infested or infected materials such as seeds or tools. Predisposing factors include cool, wet weather and high relative humidity, and large, densely planted crops of tomato. Disease development is favoured by constant cool temperatures between 10°C–24°C and a relative humidity over 75% for at least 48 hours or is at least 90% for 10 hours each day for the next 8 days, infection will take place and a late blight outbreak can be expected from 2 to 3 weeks later.

Management

Cultural Methods

- Monitor plants for late blight infections in the nursery before transplanting them to the field.
- Remove and destroy diseased leaves, branches, and diseased plants.
- Do not plant tomatoes near older plantings of either tomatoes or potatoes; otherwise, spores will spread by wind or wind-blown rain.
- Destroy all cull and volunteer potatoes.
- Use spacing between plants that allow air movement through the planting so that foliage dries as quickly as possible after morning dews or rains.
- If using own seed and the crop from which the seed was taken had late blight, dry the seed for 3 days at 22°C.
- Do not use overhead irrigation. It favours sporulation, spread, and infection.
- Remove a few branches from the lower part of the plants to allow better airflow at the base.
- Intercrop tomatoes with non-hosts, e.g., beans, maize, papaya, and bele; this will increase the spacing between plants, and reduce the spread of spores.
- Stake the plants to keep the foliage off the ground and will improve air movement around the plants.
- Remove self-grown tomatoes, potatoes, and Solanum weeds (i.e. volunteer plants) as they may have late blight infections.
- Keep up to date on late blight forecasts.
- Quickly destroy hot spots of late blight.

Host Resistance

- Grow resistant/tolerant varieties such as Arka Abhed, etc.

Chemical Control

- **Seed treatment:** Treat seed with Thiram @ 2.5 g/kg seed before sowing.
- **Foliar spray:** Spray anyone of the following fungicides as per recommended dose given below:

Fungicides	Recommended Rate
Mancozeb (63%) + Carbendazim (12%) WP	1.0 g per litre of water
Ridomil MZ72	1.0 g per litre of water
Copper oxychloride 50% WP	1000 g in 300–400 litres of water per acre
Azoxystrobin 23% SC	200 ml in 200 litres of water per acre
Captan 50% WP	1000 g in 300–400 litres of water per acre
Captan 75% WP	666.8 g in 400 litres of water per acre
Iprodione 50% WP	600 g in 200 litres of water per acre
Kitazin 48% EC	80 ml in 80 litres of water per acre
Mancozeb 35% SC	200 ml in 200 litres of water per acre
Mancozeb 75% WG	400 g in 200 litres of water per acre
Pyraclostrobin 20% WG	150–200 g in 200 litres of water per acre
Zineb 75% WP	600–800 g in 300–400 litres of water per acre
Ziram 80% WP	600–800 g in 300–400 litres of water per acre
Famoxadone 16.6% + Cymoxanil 22.1% SC	200 ml in 200 litres of water per acre
Metiram 55% + Pyraclostrobin 5% WG	600–700 g in 200 litres of water per acre
Metriam 70% WG	1000 g in 200–300 litres of water per acre

18.5 BUCKEYE ROT

The disease is also known as soft rot or Phytophthora root and fruit rot. Buckeye fruit rot of tomato is reported in all parts of the world with high relative humidity, abundant soil moisture, and warm weather. In India, it was first reported from Coimbatore by Ramakrishnan and Soumini (1947). Buckeye rot incidence may reach upto 90% under high humidity and good rainfall. Losses may range from 18%–40% due to buckeye rot. The disease incidence of upto 65% due to fruit rot disease in ripe tomatoes has been recorded from Himachal Pradesh. The disease induces fruit rot of tomato, pepper, and eggplant. The pathogen also causes damping off, foot rot, root rot, leaf blight, petiole and stem rot and fruit rot (buckeye).

Symptoms

Infection first appears on lower immature green fruits which are either on or near the soil level. Either green or ripe fruit may be infected through the uninjured skin. Infected lesions have dark brown center surrounded by water-soaked zonations. Lesions enlarge rapidly and within 3–4 days, dark brown, soft, circular or irregularly oblong lesions can cover more than half of the fruit or the entire fruit surface. As the spots enlarge, the surface of lesions develops a pattern of concentric rings of narrow, dark brown, and wide light brown bands which give a characteristic bull's eye pattern. The lesions remain firm and smooth, although the internal rotted tissue turns mushy. The margins of the lesions are smooth but not sharply defined. Initially, the infected fruit is firm, but later becomes soft and rotted. In warm and humid weather, white flocculent superficial growth of the pathogen develops profusely on lesions of the diseased fruits. The buckeye-rot pathogen may also cause a pale to dark brown, girdling stem canker which may be somewhat sunken. Affected plants usually wilt and die. In very wet weather a sparse white mildew appears on the infected stems. Fruit symptoms caused by buckeye fruit rot can be confused with symptoms of "late blight", which is caused by the fungus *Phytophthora infestans*. Lesions of buckeye rot resemble those of late blight, except that the former remain firm and smooth, whereas late blight lesions become rough and are slightly sunken at the margins. Buckeye rot lesions may cover a small part or more of the fruit.

Causal Organism and Etiology

Buckeye rot of tomato is caused by *Phytophthora nicotianae* (previously, *Phytophthora nicotianae* var. *parasitica*). The sporangia of *Phytophthora nicotianae* vary greatly and can be ellipsoidal, ovoid, pyriform, or spherical with distinct papilla and are non-deciduous. Chlamydospores are smooth, globose, and slightly yellowish with thick brown walls (20 to 60 µm in diameter).

Host range: Beans, cowpea, corn, eggplant, rhubarb, onions, okra, brinjal, chilli peppers, potato, turnip, chilli, parsley, citrus, coconut, pineapple, grape, guava, carnation, chrysanthemum, hibiscus, sesame, melons, bitter gourd, cucumber, pointed gourd, ridge gourd, and snake gourd, pumpkin, and squash.

Disease Cycle and Epidemiology

Phytophthora nicotinae is an Oomycetes, soil inhabitant that infects many Solanaceous crops and can persist in soils as oospores and chlamydospores for extended periods. The buckeye rot pathogen may also be introduced through infected seeds or transplants, by contact with infested soil or through plants from the previous crop. The pathogen is also internally seedborne and may survive on other hosts. Infected seeds give rise to infected plants. Oospores and chlamydospores germinate and form sporangia and germ tubes, respectively and cause infections. Spores can germinate in soil or on decaying debris. During wet periods, zoospores are released from the sporangia and are readily splattered by rain from the soil to the fruit. The motile zoospores swim through a film of water, encyst, and infect the fruit. The pathogen primarily infects tomato fruit lying on or near moist soil. Visible symptoms of the disease may appear within 24 hours and fruit rot develops rapidly. The mycelium on fruit surface form sporangia that causes secondary infections. The sporangia are spread by surface or irrigation water and splashing rains, farm equipment and tools, and workers. The disease development is favoured by warm, wet conditions, high rainfall, relative humidity 60%, and optimum temperature ranging from 20°C–30°C. Fruit with latent infections may decay during transit and storage if the temperature is 21°C or above.

Management

Cultural Methods

- Use certified disease-free seeds for planting.
- Stake and remove the fruits and leaves touching the ground (up to 30 cm).
- Provision of good soil drainage.
- Clipping of lower leaves and mulching.
- Grow tomatoes on the raised beds in well-drained soil. Avoid low-lying areas where drainage is poor.

- Avoid frequent irrigation using overhead irrigation as it helps spread the disease. Use furrow irrigation instead.
- Avoid frequent irrigations that keep the ground wet.
- Remove infected fruit as soon as they are seen, and before sporulations.
- Spread coconut fronds or use another mulch to prevent heavy rains from splashing soil onto low hanging fruit.
- Follow crop rotations of at least 4 years with nonhost crops, because *Phytophthora* produces thick-walled resting spores ("oospores").
- Applying mulch keeps the fruit off the soil, which reduces the chances of the disease getting into the fruit.

Host Resistance

- Use resistant varieties.

Biological Control

- Treat the seed with *Trichoderma* spp. @ 10 g per kg of seeds.

Chemical Control

- Spray Mancozeb 75% WP @ 600–800 g in 300 litres of water/acre or Propineb 70% WP @ 120 g in 40 litres of water/acre.

18.6 ROOT KNOT

Plant-parasitic nematodes are a great threat to agriculture, causing an estimated annual yield loss of over $100 billion worldwide. Among plant parasitic nematodes, the most yield-limiting group is root-knot nematodes (*Meloidogyne* spp.). Root-knot nematodes affect almost every crop in the world. Reports on *Meloidogyne* spp. infecting tomato plants dates back to the end of the 19th century. Root-knot nematodes (RKNs) are notorious plant-parasitic nematodes first recorded in 1855 in cucumber plants.

Symptoms

Root-knot nematodes (*Meloidogyne* spp.) are obligate, sedentary endoparasites that can infect both the below and above-ground parts of many crops at different developmental stages.

Above-ground symptoms: Foliar symptoms are not specific and may not always be present when nematode populations are low. Above-ground symptoms, when present, are often associated with high nematode populations, and may include stunting, chlorosis (yellowing), reduced unthrifty growth, undersized and fewer fruits, wilting of plants despite adequate soil water content, and collapse of individual plants.

Below-ground symptoms: Typical diagnostic symptoms on roots consist of characteristic galls or knots or swellings as a result of the nematode-induced expansion of root cells (Figure 18.4). The galls may vary in size and shape depending on the species, population density, susceptibility, and age of the crop. Galls caused by *Meloidogyne hapla* are much smaller than those caused by other species. The presence of galls in roots affects the absorption and transport of water and nutrients from the soil. Generally, in the initial stage of plant growth, galls (primary galls) are small. But as the nematode completes one life cycle, reinfection by second-generation J2 leads to the development of more galls, the adjacent galls coalesce to form bigger compound galls, which are easily visible at later stages of crop growth. Severely infected plants produce fewer roots and usually die rapidly. Instead of growing longitudinally, the roots grow axially due to hyperplasia and hypertrophy of cortical parenchyma cells which results in the formation of swellings (root galls or knots).

Figure 18.4 Root Knot of Tomato.

Causal Organism and Etiology

Root-knot of tomato is caused by *Meloidogyne* species (*Meloidogyne incognita*, *M. javanica*, *M. arenaria* and *M. hapla*, as well as a few emerging species such as *M. enterolobii* and *M. chitwoodi*). Root-knot nematodes (*Meloidogyne* species) are ubiquitous, obligate, biotrophic plant-endoparasites. Among the *Meloidogyne* species, *M. incognita* is the most important which causes significant damage to various economically important crops, including tomatoes. The name *Meloidogyne* means apple-shaped female. Sedentary adult root-knot nematode females are pearly white with a rounded to pear-shaped body and a protruding, sometimes bending, neck. A cyst stage is not present. The cuticle annulations in full-grown females are only visible in the head region and the posterior part where a characteristic unique cuticular pattern or perineal pattern can be observed around the perineum (vulva–anus region). The non-sedentary male is vermiform, clearly annulated, and ranges in length from 600 to 2500 μm. The shell of eggs has three layers, the outer vitelline, middle chitinous, and inner lipid layer. The J3 and J4 retain old cuticles, the pointed tails of J2 still visible, also called spike-tailed stages. Adult females lay about 500–1,000 eggs in a gelatinous matrix which is secreted on the root surface. The eggs are ellipsoid. About 400–500 eggs are present in the sac but the number may go upto 1000 or even more. Each female may lay 30–80 eggs per day, the number depending on the host plants and the environmental conditions.

Host range: About 100 species of *Meloidogyne* are known to attack more than 3,000 plant species. Tomato is often referred to as a universal host for *Meloidogyne* species.

Disease Cycle and Epidemiology

Root-knot nematodes evade plant immunity, hijack the plant cell cycle, and metabolism to modify healthy cells into giant cells (feeding sites). The life cycle of root-knot nematodes completes in three basic stages: egg, juvenile (J), and adult stages. The life cycle of *Meloidogyne* species starts from the eggs which are deposited by the females within the host tissues or outside on the surface of galled roots in a gelatinous matrix that protects eggs from extreme environments (e.g. high temperature) and microbes. In absence of the hosts, eggs represent the survival stage of the nematodes in the soil. Root-knot nematodes do not require any specific stimulus from the host and the eggs hatch freely in the water. Under favourable conditions, egg development takes place and the first stage juvenile (J1) is formed. First juveniles (J1) molt while still within the egg-shell and become second-stage juveniles (J2) or pre-parasitic. Under suitable conditions, J2 hatches out, moves freely in the soil in search of new roots of the same plant or some other plant, and is the only infective stage. Under adverse conditions, J2 can survive in an inactive anhydrobiotic phase when the body shrink and coils. In root-knot nematodes, the genes related to the parasitism express in cells of esophageal glands and encode enzymes that are secreted into the root cells. At the onset of parasitism, second-stage juveniles (J2) penetrate the roots by repeated stylet thrusts and/or enzymes secreted by the oesophageal glands and move toward the vascular system of root tissues at this stage J2 becomes sessile where they initiate the development of permanent feeding sites termed "giant cells". This feeding site serves as a source of water and nutrients for the growth of J2 and other development stages. Giant cells (2 to 12, usually about 6), feeding sites for the root-knot nematode, are established in the phloem or adjacent parenchyma. J2 becomes swollen (now called parasitic J2) after the beginning of the feeding process. Under optimum conditions, J2 molts to become J3, which develops quickly into the J4 stage after third molt, and then adult females lay eggs in a gelatinous sac. The eggs finally reach the soil and repeat the life cycle. The whole life cycle completes in about 25 days at 25°C to 30°C, which is optimum for most species. During the winter season under north Indian conditions, the life cycle may extend to 60–80 days depending on prevailing temperatures. Very little activity of any *Meloidogyne* species occurs above 40°C and below 5°C. High temperature (40°C–50°C) kills nematode larvae. Thus 7–8 overlapping generations are completed in a year. In tropical countries, the most distributed species are *M. incognita* and *M. javanica*; in cold countries, with a temperature of less than 15°C, *M. hapla* is the most distributed. The development of nematodes occurs in the range of 13°C–34°C, with optimal development at 29°C. If the soil temperature is below 10°C, it cannot develop, and its damage starts at 15°C. In addition, root infection of root-knot nematodes breaks down the resistance and increases the susceptibility of the host plant to other pathogens, such as *Fusarium oxysporum*, etc.

Management

Cultural Methods

- Only seedlings with roots free of galls should be selected for transplanting.
- Two or three deep summer ploughing at 10–15 days intervals during may/June are very effective in killing most of the J2 in soil due to desiccation and exposure to high temperatures.
- Follow crop rotation with non-hosts or resistant crops or poor hosts or antagonistic hosts. Cereal crops like wheat, maize, rice, sorghum, pearl millet or antagonistic crops like onion, garlic, and sesame can be used for rotations/sequences with susceptible crops. By alternating to a crop that the nematodes can neither feed on nor complete their life cycle on, can reduce the root-knot nematode populations in the soil. However, crop rotation is not of much use due to the wide host range of the root knot nematodes.
- Incorporation of cruciferous green manures and roots of marigolds (*Tagetes* spp.) to the soil helps in reducing the nematode populations.
- Summer fallowing is a cheap and effective way to reduce nematode population. This will not stop nematode hatching from the eggs, but in absence of host plants, the young worms will die.
- Remove the weeds regularly.
- Soil solarization with clear plastic tarps during summer. Cover the moist soil with polythene sheets in small areas like nursery beds for 3–4 weeks before sowing. The increased soil temperature helps to kill RKNs and many other soilborne pests and pathogens. In moist beds, nematodes will hatch out from eggs, move around for roots, and will die of starvation.
- Intercropping of antagonistic crops like *Tagetes* in between susceptible crops or cultivations of onion and garlic to reduce nematode populations.
- Rabbing (a process of sterilization) is another method to raise root knot nematode free seedlings. After ploughing the nursery area to a fine tilth, a 15–20 cm thick layer of rice husk (20 kg/m^2) is spread on the surface and burned. The ash is mixed with the top soil and a week later seeds can be sown.
- Destroy galled roots after harvesting.
- Keep farm tools soil-free.
- Organic amendments enhance biological suppression of plant parasitic nematodes in the soil. The organic matters mostly used to manage root knot nematodes are poultry manure, pigeon litter, sawdust, and various crop residues.

Host Resistance

- The use of *Meloidogyne*-resistant cultivars has been an effective management tool for root knot nematode; however, not many resistant cultivars are commercially available, and resistance may be overcome by new emerging RKN species such as *M. enterolobii.*
- In modern tomato cultivars *Mi* genes derived from *Lycopersicon peruvianum* L. are conferring resistance to the major species of *Meloidogyne*. Resistance in tomatoes to

RKNs was first observed by Bailey 1941 in the wild species (*Lycopersicon peruvianum* L.) The major problem is that their resistance breaks down at high temperatures. The resistance mediated by *Mi-1* genes is lost at high temperature above (28°C). The gene *Mi-1* is the most commonly used and the only resistance gene commercially available gene. The tomato gene *Mi1.2* confers resistance not only to root knot but also to aphids (*Macrosiphum euphorbiae*) and whitefly (*Bemisia tabaci*).

- Tomato varieties like Pusa Ruby, S-120, NTDR-1, VFN-8, Nematex, PNR 7, Pusa 120, A-1-1-2, Punjab 6NR-7, VFN-8, VFN-Bush, SL-12, Resistant Bangalore, VFX-360, Hisar Lalit, Karnataka hybrid, Mangla hybrid, etc. are resistant to root knot nematode.
- Grafting a desirable scion onto a root-knot nematode resistant rootstock may be a helpful way to avoid crop damage and yield loss due to root-knot nematodes.

Biological Control

- *Paecilomyces lilacinus* (an oviparasitic fungus) *Pasteuria penetrans* (mycelial, endospore-forming bacterial parasite), *Pseudomonas fluorescens* and *Verticillium chlamydosporium*, are promising biocontrol agents of root knot nematodes.
- Apply *Pseudomonas fluorescens* @ 10 g/m^2 in nursery.
- Biocontrol products have been shown to reduce root galling, egg masses, and overall nematode populations.

Chemical Control

- Treat the nursery beds with the application of Carbofuran @ 3 kg a.i./ha (0.3 g a.i. per m^2).
- Dip seedling roots in Thionazin (500 ppm solution) for 15 min. or Carbosulfan (1000 ppm solution) for one hour.

18.7 VIRAL DISEASES

18.7.1 Tomato Yellow Leaf Curl

The most severe viral disease of tomatoes is the notorious tomato yellow leaf curl disease (TYLCD) which causes substantial yield losses in many parts of the world including India. Tomato yellow leaf curl is also known as "Curly Top" disease of tomato. TYLCD was first observed in Israel in 1939–1940 and reported by Cohen and Harpaz in 1964. The virus induces severe yield losses which, depending on the age of the plant at the time of infection, can reach 100%. In India, it is a very serious disease of tomatoes, especially in summer and rainy season crops.

Symptoms

Leaf symptoms include mosaic, intervenal yellowing, vein clearing, and crinkling and puckering accompanied more often by inward rolling, mottling, chlorosis of leaf margins, smaller, and deformed leaflets (Figure 18.5). The older leaves become leathery, and brittle. The disease also induces severe stunting, short internodes, erect shoots, bushy growth, and partial or complete sterility depending on the stage of the crop. Flower drop (abscission) is common. Infected plants bear few or no fruits. Infected plants are generally distributed randomly or cluster in the field.

Figure 18.5 Leaf Curl of Tomato.

Causal Organism and Etiology

Tomato yellow leaf curl disease is caused by *Tomato yellow leaf curl virus*. It is a species in the genus Begomovirus and the family Geminiviridae. TYLCV is a monopartite DNA virus with the circular genome that contains six genes with two genes on the viral strand (V1–V2) and four genes on the complementary sense strand (C1–C4). *Tomato yellow leaf curl virus* (TYLCV) is a tomato (*Solanum lycopersicum*)-infecting plant virus transmitted by whitefly (*Bemisia tabaci*). It belongs to the genus *Begomovirus* of the family *Geminiviridae* and has a single-stranded circular DNA genome of about 2.8 kb encapsidated in a twinned icosahedral virion. Like all members of the Geminiviridae, TYLCV has a twinned (geminate) particle that is 18–20 nm in diameter and 30 nm long, apparently consisting of two incomplete icosahedral fused together in a structure with 22 pentameric capsomers and 110 identical protein subunits. The viral genome is comprised of a closed circular single-stranded (ss) DNA of nearly 2.8 kb in size. The first begomovirus satellite discovered, referred to as tomato leaf curl virus-sat (ToLCV-sat), was identified in tomato plants infected with the monopartite begomovirus tomato leaf curl virus (ToLCV) originating from Australia. Most monopartite begomoviruses are associated with a group of ssDNA satellites collectively known as betasatellites. Unlike most whitefly transmitted geminiviruses whose genomes are bipartite, the genomes of TYLCVs (with the exception of *Tomato yellow leaf curl Thailand virus* and *Tomato yellow leaf curl Kanchanaburi virus*, which possess bipartite genomes) are monopartite, i.e. they contain only one genomic circular single-stranded DNA (ssDNA) molecule of about 2.8 kb.

Host range: In addition to tomato, TYLCV can also naturally infect other crops including pepper (*Capsicum* species), common bean (*Phaseolus vulgaris*), cucurbit (*Cucumis* species) and eustoma (*Eustoma grandiflora*) have been reported to be TYLCV hosts. Chilli pepper (*C. chinense*), tobacco (*Nicotiana tabacum*), petunia (*Petunia* × *hybrida*), and common weeds (*Solanum nigrum* and *Datura stramonium*). TYLCVs also infect the experimental host *Nicotiana benthamiana*.

Disease Cycle and Epidemiology

So far only a few weeds, the common bean, and pepper are known to act as the source of the virus depending upon the isolate and vector biotype. Natural infection of annual weed, *Datura stramonium*, *Solanum nigrum*, *Solanum luteum*, and *Euphorbia* sp. is by TYLCV-Sar. TYLCV is also capable of establishing symptomless infections in other hosts, such as peppers and cucurbits. These plants serve as reservoirs for the virus and despite the lack of symptom development, the whitefly vector (*Bemisia tabaci*) is capable of acquiring and transmitting TYLCV to other plants, including tomato. The virus is phloem limited in its hosts, and it is transmitted exclusively by the sweet-potato whitefly (*Bemisia tabaci*), in a persistent and circulative manner. The virus is retained throughout the life of the adult insect after acquisition and moves through the insect body to the salivary glands where it can leave the whitefly body via its saliva. Female *B. tabaci* are more efficient virus transmitters than males. The virus cannot be spread in any other way.

Management

Cultural Methods

- Use seedlings for planting from a nursery that is fully protected seedlings from early infestations of whitefly.
- Use virus- and whitefly-free transplants.
- Avoid planting new fields near older fields with TYLCV-infected plants.
- Cover plants with floating row covers of fine mesh (Agryl or Agribon) to protect from whitefly infestations.
- Rogue out the virus-affected plants as soon as the symptoms are observed.
- Remove and destroy old crop residue and volunteers on a regional basis.

Host Resistance

- The best long-term management is to use resistant varieties such as Arka Ananya, Kashi Vishesh, Kashi Amrit, COTH 2, and TNAU Tomato Hybrid Co3.

Management of Vectors (Whiteflies)

Biological Control

- Spray neem seed kernel extract 4% (NSKE) or neem oil @ 8–10 ml/L or neem soap @ 10 g/L.
- Parasitoid *Eucarsia porteri* and *E. sophi* have been found effective against whitefly biotype B, but needed two different host plants for the males and females to develop.
- *Beauveria bassiana*, an entomogenous fungus, at 1×10^7 spores/ml in 1% groundnut oil or coconut oil inhibited infectivity of the three instar stages up to 86%, and mortality of adults up to 75.6% in 24 hours and up to 100% in 72 hours on cotton.
- Blue, red, and violet are inhibitory to whiteflies, and yellow plastic mulch protects tomato against TYLCV.
- Install yellow sticky traps in the field @ 25–30 traps/hectare after transplanting.

Chemical Control

- If the incidence is high, spray Dimethoate 30% EC @ 1 ml/L or Cypermethrin (0.01%) at 10–15 days intervals.
- A fortnight (15 days) after transplanting, give need-based sprays of Thiamethoxam 25% WP @ 0.4 g/L or Dimethoate 30% EC @ 1 ml/L or Lastraw @ 5 ml/L (Apply 2–3 times at weekly intervals for effective control). Do not repeat at the fruiting stage as this may leave harmful residues in fruits.

18.7.2 Tomato Mosaics

Tomato mosaic can be caused by many different viruses such as *Tobacco mosaic virus* (TMV), *Cucumber mosaic virus* (CMV), *Potato virus X* (PVX), and *Potato virus Y* (PVY). The common tomato mosaic was long considered to be caused by a strain of TMV but it is now considered to be a distinct viral species (*Tomato mosaic virus*, ToMV) of the same tobamovirus group, also transmitted by contact. Viruses of this group cause significant yield losses in tomato. ToMV is widely present in every continent and found more frequently than TMV on tomato, and pepper, both in the field and under protected cultivation. Mixed infection especially of CMV and PVY is also very common in tomato and causes more serious damage. TMV and ToMV are highly infectious, extremely persistent, and easily transmitted during the handling of plants.

Symptoms

***Tomato mosaic viruses* (ToMV):** Whitening of the leaf, mottling, leaflets showing green, yellow, or white patches of mosaic. The leaflets may also have raised tissue, be variably crinkled, wrinkled, and sometimes, even filiform, when plants lack light. Chlorotic rings are visible on green or mature fruit. Internal brown necrotic lesions are also found (internal browning). A co-infection ToMV-PVX leads to the appearance of dark brown and black streaking on the petiole and stem, a very damaging tomato syndrome, called "double streak".

***Tobacco mosaic virus* (TMV):** Slight mottling, mosaics on leaflets. Leaflets may be filiform in periods of low light. The petals and sepals are sometimes deformed; there is also a decrease in the number of pollen sacs. This virus is often mistaken for ToMV which is better adapted to tomato.

***Potato virus Y* (PVY):** Symptoms vary depending on the strains: Common strains causes discrete mottling and green mosaic patterns, diffused chlorotic spots, some distortion of leaves, vein banding (dark green bands form along leaf veins), leaf curling, and bushy appearance of plants. Necrotic strains cause reddish brown, necrotic interveinal lesions on leaves and longitudinal necrotic streaks on the petioles and stem. This virus causes no symptoms in tomato fruit.

***Cucumber mosaic virus* (CMV):** Symptoms on tomato may vary depending on the virus strain, severity, time of infection, and climatic conditions. However, three types of symptoms are mainly observed: (i) mottling, mosaic on young leaflets, bushy and considerably stunted plants; (ii) deformation of the leaflets, extremely filiform, which have the appearance of a "shoelace" or "shoestring" or "fernleaf", this is most characteristic feature CMV infection (iii) spots, necrotic lesions appear on leaflets and stem, extending to the apex of the plant. In addition, severe necrotic brown, raised, round lesions cover large areas of the fruit.

Causal Agent and Etiology

Tomato mosaic disease is caused by ToMV, TMV, PVY, and CMV.

- The virus causing the common mosaic of tomato is known as *Tomato mosaic virus* (ToMV).
- TMV and ToMV are viruses in the tobamovirus group. These viruses have straight rod particles that measure 300 × 18 nm and contain single-stranded, linear RNA genomes.
- ToMV has different strains such as tomato aucuba mosaic, tomato emanation mosaic, yellow ring spot strain, and tomato rosette strain which produce different symptoms and often described as different diseases.
- PVY is a potyvirus with long (730 × 11 nm) flexuous rods that contain single-stranded RNA.
- CMV is a cucumovirus in the family *Bromoviridae*, with isometric particles (29 nm in diameter) and contains three ssRNAs.

Disease Cycle and Epidemiology

TMV and *ToMV*

TMV is transmitted by contact, almost never by seeds. ToMV is transmitted by contact and by seeds. TMV has no known vector and is readily transmitted mechanically. ToMV also lacks a known vector, is transmitted mechanically, and can be seedborne in tomato. TMV and ToMV are two of only a few plant viruses that are not transmitted by insects. In contrast to many other plant viruses, TMV and ToMV can survive for up to 50 years in plant debris and for weeks to months on trellises or wooden stakes. Both viruses can infect tomato and other solanaceous plants. Because ToMV infection is aggressive and highly contagious, many breeding programs were started to find sources of resistance against this virus. Several tobamoviruses (e.g. *Tobacco mosaic virus* and *Tomato mosaic virus*) are seed-borne, which contributes to disease spread. The seed transmission of ToMV and TMV has been reported from seed to seedlings to be 1–13% and 1–10%, respectively.

PVY

PVY is vectored in a nonpersistent manner by several aphids, with *Myzus persicae* being particularly important. This virus is also readily transmitted mechanically. Transmitted by about 40 different aphids in a nonpersistent mode (including *M. persicae, Aphis gossypii, A. fabae, Acyrthosiphon pisum, Macrosiphum euphorbiae*).

CMV

It infects over 1200 species in over 100 families of monocots and dicots, including many vegetables, ornamentals and woody and semi-woody plants. Because of its wide host range, numerous weeds can serve as reservoirs for CMV and contribute to the virus spread to crops at the beginning of the season. *Cucumber mosaic virus* on tomato is not seed transmitted and typically enters the field or gardens through infected transplants or from weed hosts of the virus.

Transmitted by aphids in a nonpersistent manner. Over 90 species are capable of acquiring and transmitting the virus, including: *Myzus persicae*, *Aphis gossypii*, *A. craccivora*, *A. fabae*, *Acyrthosiphon pisum*. In tomatoes, *Cucumber mosaic virus* is much less prevalent than *Tobacco/ Tomato mosaic virus*.

Management

Tobacco and *Tomato mosaic viruses* are difficult to control, as they can survive harsh conditions for many years. CMV is also very difficult to control because of its extremely wide host range. Once a plant is infected, there is no cure.

Cultural Methods

- Use of virus-free seedlings. It is the most important step for the management of tomato mosaic disease. Because ToMV is seedborne in tomatoes, use seed that does not have significant levels of the pathogen.
- Choose only transplants showing no clear symptoms while purchasing from any nursery.
- Soak seeds in 10% Trisodium phosphate (Na_3PO_4) solution for at least 15 minutes to eliminate externally located virus on the seed coat.
- Treat freshly harvested seed with dilute Hydrochloric acid (HCl) that will inactivate virus contaminating seed coats.
- To produce healthy seedlings the seedbeds should be those in which no Solanaceous crop susceptible to TMV/ToMV had been grown for the last 4–6 months.
- The cultivation of plants can be done in sterilized compost in plastic bags ('growbags') or hydroponic systems in which the nutrient solution is virus-free. Virus-free nutrient solutions can be produced by pasteurization.
- Use new potting soil, pots, and string every time.
- To inhibit the mechanical transfer of the virus from infected to healthy plants spray milk.
- Disinfect all tools by dipping them in 3% (w/v) tri-sodium phosphate solution or dipping contaminated tools in 20% powdered milk solution for one minute will kill the virus.
- Nursery workers should thoroughly wash their hands and any contaminated clothing.
- Remove symptomatic plants that may slow the spread of disease.
- Avoid planting in fields where tomato root debris is present, as the virus can survive long-term in roots.
- Wash hands with soap and water before and during the handling of plants to reduce the potential spread between plants.
- Disinfect tools regularly–ideally between each plant, as plants can be infected before showing obvious symptoms.
- Soaking tools for 1 minute in a 1:9 dilution of germicidal bleach is highly effective or a 1-minute soak in a 20% weight/volume solution of nonfat dry milk and water is also very effective.

- Avoid using tobacco products around tomato plants, and wash hands after using tobacco products and before working with the plants. Tobacco in cigars, cigarettes, and pipe tobacco are an important source of ToMV.
- Tobacco in cigarettes and other tobacco products may be infected with either ToMV or TMV, both of which could spread to the tomato plants.
- Scout plants regularly. If plants displaying symptoms of ToMV or TMV are found, remove the entire plant (including roots) and burn infected plants.
- Do not use infected plant material for composting due to the longevity of the virus.
- After working with diseased plants, thoroughly disinfect all tools and hands as outlined above.
- Avoid working with healthy plants after working in an area with diseased plants.
- At the end of the season, burn all plants from diseased areas, even healthy-appearing ones, or bury them away from vegetable production areas.
- Disinfect stakes, ties, wires or any other equipment between growing seasons using the methods noted above.

Physical Control

- Treat the seed with hot water at 50°C for 25 minutes.
- Seed infested with ToMV can also be treated with dry heat (70°C for 2 to 4 days) or with Trisodium phosphate (10% for 15 minutes) to eliminate the virus present externally on seed coats and in the endosperm.
- Soil sterilization by heat is also recommended.

Host Resistance

- Use resistant varieties. Three tomato genes Tm-1, Tm-2, and Tm-2^2 mediate resistance against tobamoviruses including *Tomato mosaic virus* (ToMV) and *Tobacco mosaic virus* (TMV). In contrast to Tm-1 and Tm-2, Tm-2^2 mediates much more durable resistance and has been used in most commercial tomato cultivars against infection with tomato mosaic virus (ToMV) for several decades worldwide.

Management of Aphid Vectors

- Apply a recommended dose of nitrogenous fertilizers.
- Use yellow sticky traps for aphids @ 25–30 traps/hectare.
- Release first instar larvae of green lacewing @ 4000/acre.
- Conserve predators such as ladybird beetles (*Coccinella septumpunctata* and *Menochilus sexmaculata*) and parasitoids such as *Aphidius colemani* etc.
- Spray Neem seed kernel extract 4% (NSKE) or Neem oil @ 8–10 ml/L or Neem soap @ 10 g/L.
- Spray Dimethoate 30% EC @ 264 ml in 200–400 litres of water/acre or Cyantraniliprole 10.26% OD @ 360 ml in 200 litres of water/acre.

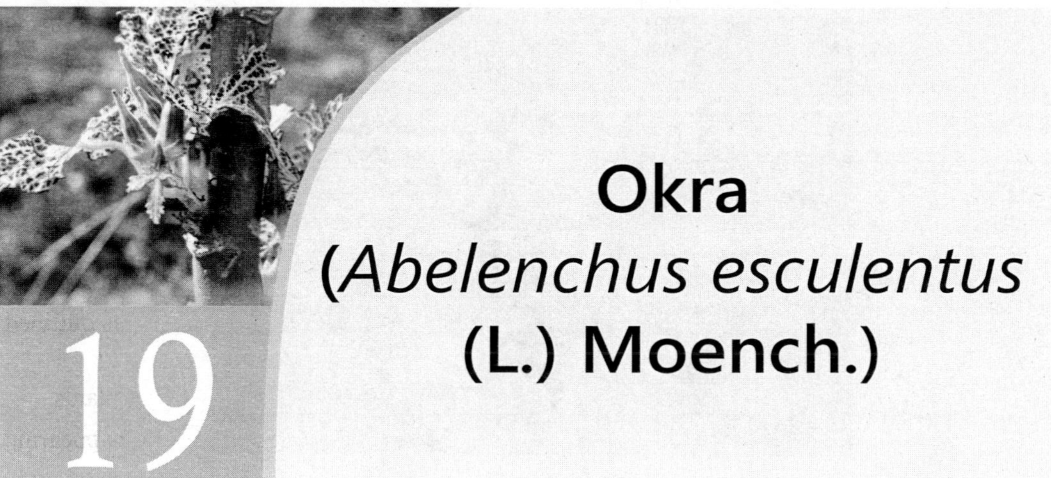

19 Okra (*Abelenchus esculentus* (L.) Moench.)

19.1 YELLOW VEIN MOSAIC

Yellow vein mosaic disease (YVMD) is a major biotic constraint on okra. It is very destructive to the okra crop and causes 50%–95% yield losses in India. It was first reported by Kulkarni in 1924 in Maharashtra (India). This was first described as yellow vein banding. It is characterised by different degrees of chlorosis and yellowing of veins and veinlets, smaller leaves, fewer and smaller fruits, and stunting.

Symptoms

On leaves: A homogenous interwoven network (due to vein clearing and veinal chlorosis) of yellow veins enclosing islands of green tissues (Figure 19.1A). The yellow network and thickening of veins and veinlets are very prominent symptoms (as a result of chlorosis). Enations (raised structures) are observed on the under surface of the infected leaf.

On fruits: Reduced in size, exhibit pale yellow colour, malformation, small and tough in texture. In severely infected ratoon crops, yellow and green strip-patterns develop on fruits (Figure 19.1A).

On the whole plant: Stunting and yellowing (because of severe infection, the chlorosis of the interveinal area occur which results in the yellowing of entire leaves).

Causal Agent and Etiology

Yellow vein mosaic of okra is caused by *Bendi yellow vein mosaic virus* (BYVMV). This virus is known as geminivirus (i.e. having paired or twin genome (ssDNA molecule) particle encapsidated in two virions) and measures 18 × 30 nm.

Figure 19.1 Yellow Mosaic of Okra. (A) Yellow vein mosaic symptom on leaves and yellow strips on young fruits, (B) Whiteflies feeding on lower surface of okra leaf.

Disease Cycle and Epidemiology

In absence of the host crop, BYVMV survives in weed hosts such as *Croton sparciflorus*, *Malvastrum tricuspidatum*, and *Ageratum* spp. which serve as a reservoir of primary inoculum for infection. BYVMV transmitted by white fly (*Bimisia tabaci*) (Figure 19.1B). It transmits the virus by a non-persistent, circulative, and nonpropagative manner. The vector retains the virus during successive molting from the egg to the adult stage. Warm humid weather favours the vector population. It is neither a sap nor seed and pollen transmissible virus. Under experimental conditions, it has also been transmitted by grafting.

Management

Cultural Methods

- Field sanitation, rouging, and destruction of virus-affected plants.
- Plant tall border crops like maize, sorghum or pearl millet to reduce whitefly infestations (4 rows).

- Install yellow sticky traps @ 10 to 20 per acre.
- Peppermint plants act as a repellent for whitefly.
- During the summer season, avoid sowing the susceptible varieties when the whitefly activity is high.

Host Resistance

- Plant disease resistance/tolerant varieties to reduce disease incidence, e.g. Parbhani Kranti, Janardhan, Haritha, Pusa Sawani, Pusa A 4, Arka Abhay, Arka Anamika, Varsha Uphar, Hisar Unnat, Hisar Naveen, HBH-142 (F1 hybrid), Gujarat Anand Okra-5, CO 1, CO 3, COBhH 1, Azad Bhindi-1, Azad Bhindi-3, Punjab-7, and Hybrid-6.

Chemical Control

- Soil application of Carbofuran (1 kg a.i./ha) at the time of sowing and 4–5 foliar sprays of Dimethoate (0.05%) or Metasystox (0.02%) or Nuvacron (0.05%) at an interval of 10 days effectively controls the whitefly population.
- Control vector population by application of Chlorpyriphos 2.5 ml + Neem oil 2 ml per liter of water.

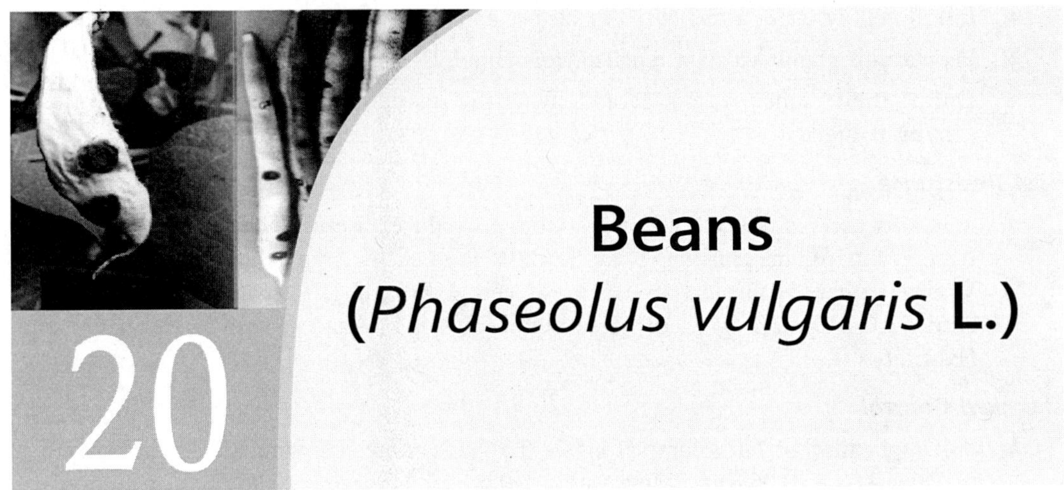

20 Beans (*Phaseolus vulgaris* L.)

20.1 ANTHRACNOSE

Anthracnose is a major disease of common beans and can also occur on other legumes. The losses can approach 100% when the severely affected seeds are sown under favourable environmental conditions. The pathogen is seed borne and planting of infected seed causes 80% to 100% yield loss. Hutchinson and Ram Ayyar (1915) noticed first time the occurrence of bean anthracnose in the Nilgiri Hills in India. In 1921, M.F. Barrus demonstrated that bean anthracnose is seedborne.

Symptoms

The most characteristic symptoms appear on cotyledons, leaves, petioles, stem, and immature pods and seeds.

Seedlings: On small seedlings, dark brown to black lesions appear on the cotyledons. Seedling stems may have rust-coloured flecks or 1/4-inch-long elliptical, sunken lesions that weaken the stems, causing stunting, or girdle the stems, causing seedling death.

Leaves: Initial symptoms appear as necrosis on the veins and adjoining tissues of the underside of the leaves. Later, elongated, angular, and brick red to purple spots appear on the leaves with veins, especially on the under-leaf surface showing dark streaks. As the disease progresses, they also appear on upper leaf surfaces.

Stems and petioles: Dark brown, sunken, circular to elongate or elliptical, 1–2 cm diameter lesions that may girdle the stem. Under wet weather, pinkish to dark brown fruiting bodies (acervuli) from which pink to orange mass of spores ooze out, can be seen in the center of the lesions, especially of stems, petioles, and pods.

Pods: Circular to irregular, tan to rust-coloured, sunken cankerous lesions, with a brown or purple border on bean pods are the most noticeable symptom of anthracnose (Figure 20.1). Inside the infected pod, the seed coat may have brown to black lesions. During severe outbreaks of anthracnose, pods may dry, shrivel and fail to fill.

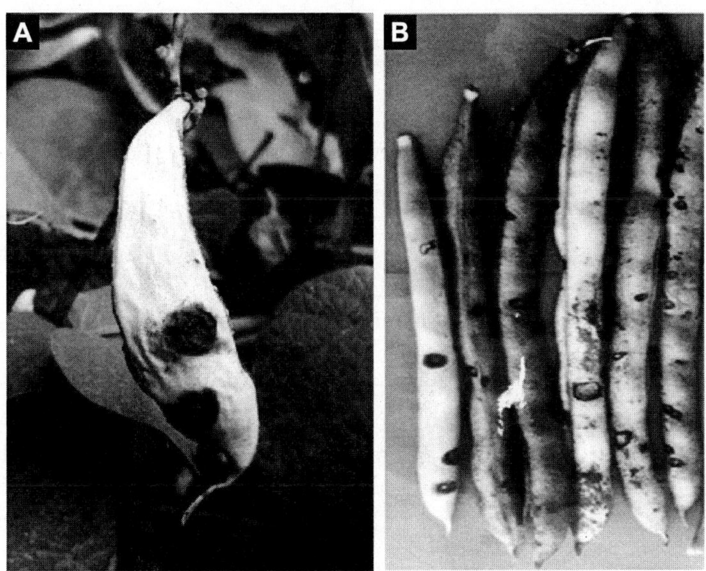

Figure 20.1 Anthracnose of Beans. (A) Broad bean pods with black, sunken lesions or reddish-brown blotches, (B) black-red sunken spots on infected pods of common bean. As these lesions become older, the edges develop a black ring with a red outer border and show a pink center containing the acervuli and conidia of the fungus.

Seed: Yellow to brown lesions on the seeds, which may be underdeveloped and produce poor germination. The seeds obtained from heavily infected pods show brown to light chocolate coloured sunken cankers on the seed coats. Seeds with less severe infection show yellowish to brown sunken lesions but are not always easy to distinguish from those cause by certain other organisms.

Causal Organism and Etiology

Bean anthracnose is caused by the fungus *Colletotrichum lindemuthianum* and is important in bean fields. Mycelium is branched, septate, hyaline at first, and dark coloured with age. Acervuli develop beneath the cuticle. Conidia are borne on short conidiophores. Setae are few, brown, and septate. Conidia are one-celled, hyaline, and cylindrical with rounded ends or with one end slightly pointed. Mycelium is branched, septate, and hyaline. Conidia are produced singly at the ends of free hyphae, or on stromatoid masses in acervuli. During the infection of bean *C. lindemuthianum* initially produces biotrophic primary hyphae that are entirely intracellular. They are followed by necrotrophic secondary hyphae that are inter- and intracellular.

Disease Cycle and Epidemiology

The pathogen overwinters in infected seeds and plant debris in the soil. In infected seeds, the pathogen mostly presents in seed coats and cotyledons, and rarely in embryonic axes and can survive for several years or as long as the seed is viable. The fungus can survive in crop debris for upto 2 years. Conidia and/or dormant mycelia in the infected seeds and/or infected plant debris germinate and infect the young seedlings. First lesion develops and produces spore masses on the cotyledons which serve as the source of secondary inoculum. Water is essential for their release from the gelatinus matrix and usually, they are washed down by rain or dew to stems or they are splashed to new leaves by raindrops falling on acervuli. The movement of insects, animals, and man may spread conidia particularly when foliage is moist. The infection process begins with the adhesion of conidia to the plant's surface. Conidia germinate on the host surface and the germ tube differentiates into a specialized penetration structure known as appressorium. Melanised appressorium produces high turgor pressure to penetrate the host surface directly. *C. lindemuthianum* is a hemibiotroph that first switches biotrophic and then necrotrophic life style. In a biotrophic lifestyle, an infection peg emerges from the appressorium and afterward the fungus forms haustorium (nutrient absorbing structure) and primary hyphae. Biotrophic hyphae spread to a few adjacent cells and then the fungus switches to necrotrophy by producing secondary hyphae. Bean anthracnose is favoured by heavy and frequent rains, cool, wet weather, moderate temperatures ranging from 13°C–26°C with an optimum of 17°C, relative humidity above 92%, and free moisture as it required for sporulation, spores germination, and initial infection.

Management

Cultural Methods

- Use disease-free seeds harvested from disease-free regions for planting.
- Remove all plant debris after harvest.
- Deep ploughing to bury debris.
- Do not plant bean seeds in an area that had a disease history for 2–3 years.
- Follow 3 years of crop rotation with non-host crops like cereals crops (rice wheat and maize) to break the disease cycle of the fungus.
- Ensure adequate plant spacing which promotes foliar drying.
- Select fields in high areas, with proper drainage and sufficient wind for aeration for planting green beans.
- Remove weeds on regular basis. This will promote proper air circulation and decrease moisture in the foliar canopy.
- Avoid sowing before the recommended planting dates, because cool conditions favour the development of anthracnose.
- Avoid overhead irrigation and avoid splashing soil onto the plants during watering. Since it will wet and liberate fungal spore masses on foliage.
- The first opportunity for the management of seed-borne diseases is to eradicate or reduce the pathogen inoculum in the seed production field.

- Disease-free seed can be produced either by surface irrigation in semi-arid regions or under a pedigreed seed programme, in which seed plots are isolated and subjected to strict inspection for disease-free seed.
- Avoid wounding bean plants.
- Avoid re-using irrigation water because bacterial pathogens can be transmitted in water.

Physical Control

- A hot-water seed treatment by soaking at 55°C–72°C for 15 hours followed by another soaking in boiling water (100°C) for 25 min kills the fungus in infested seeds without reducing germination.
- Soil solarization through covering the soil with transparent plastic sheeting for one month before sowing reduces both the severity and incidence of anthracnose.

Chemical Control

- Seed treatment with Mancozeb (@ 3 g/kg seeds) or Carbendazim (@ 2 g/kg of seed) followed by Carbendazim foliar spray @ 0.5 kg/ha.
- Fungicide sprays of fixed copper are the only recommended chemical that can be used on lima beans for anthracnose control.
- Spray the crop with Copper oxychloride (0.25%).
- Treat the seed with Benlate or Carbendazim @ 2 g/kg of seed or Mancozeb @ 3 g/kg of seeds.

20.2 BACTERIAL BLIGHTS

20.2.1 Common and Fuscous Blight

Common bacterial blight or more frequently referred to as the common blight is an important disease of common bean. The disease causes significant yield losses (up to 40%) in temperate and tropical areas and is still considered one of the major constraints in dry bean production in many countries. In 1897, E.F. Smith described the causal bacterium and named it *Bacillus phaseoli*. In India, Patel and Jindal first observed the disease at Pune, Maharashtra in 1971.

Symptoms

Seedlings: Small, water-soaked lesions may develop on cotyledons and/or primary leaves and stems of seedlings raised from infected seeds.

Leaves: Symptoms initially appear as small, water-soaked spots (lesions) on the leaves and frequently occur at the leaf margins. The lesions gradually enlarge and frequently coalesce forming large dead areas of irregular shapes, and the yellowing of leaves becomes more stabilised. Dark discoloured lesions also develop along the main vein, indicating systemic infection. Infected tissues of the lesions are flaccid, dry, turn brown, and lesions are often surrounded by a narrow, distinct, lemon-yellow halo. The dry centers of the spots may tear and fall out giving the leaves a tattered and shredded appearance. In severe cases, leaflets are killed and premature defoliation of plants occurs.

Stems and petioles: Water-soaked, occasionally sunken lesions appear on the stem, which enlarges longitudinally, and turn brown. Mostly these lesions are present near the first or cotyledonary node. The lesions frequently break open at the surface exuding yellow bacterial ooze. The water-soaked cankers or rot may girdle the stem at the cotyledonary node, and this phase is known as "joint-rot". Generally, the plants break off at this point due to heavy rain or strong wind. The presence of a sufficient amount of bacteria in the xylem tissue may cause plant wilting by plugging the vessels or disintegration of the cell wall. On petioles and pulvinus, initially, the lesions are water-soaked, but may turn brown later on.

Pods: Infected pods exhibit large, water-soaked areas somewhat circular, and slightly sunken lesions. As the disease progress, pod lesions become distinctly brick-red and more deeply pitted. Pod lesions are covered with yellow bacterial ooze that can dry to a yellowish crusty mass, under highly humid conditions.

Seed: Pod infection often causes discolouration, shriveling, reduction in size, and bacterial contamination of seeds, however, some seeds may appear symptomless. In comparison to seeds of coloured varieties, butter-yellow to brown discolouration of seeds is more prominent on white-seeded varieties.

The difference in common blight and fuscous blight symptoms

The symptoms of common and fuscous blights are similar and generally cannot be distinguished under field conditions. Though most symptoms of common and fuscous blights are similar, the intensity of seed discolouration and brown discolouration at the stem nodes is more in the latter disease. In fuscous blight, the lesions on the stem are also more inclined to split the stem. Common blight is often found in association with fuscous blight.

Causal Organism and Etiology

Common blight is caused by *Xanthomonas axonopodis* pv. *phaseoli* (also known as *X. campestris* pv. *phaseoli*). Foscous blight is caused by the variant of *X. axonopodis* pv. *phaseoli,* i.e. *Xanthomonas axonopodis pv. phaseoli var. fuscans.* Both are non-spore-forming, aerobic, gram-negative, rod-shaped, and motile with a single polar flagellum. Colonies of *Xanthomonas axonopodis* pv. *phaseoli* are yellow, convex, and slimy on glucose-containing media while *X. axonopodis* pv. *phaseoli* var. *fuscans* produces a brown pigment in culture media.

Disease Cycle and Epidemiology

The most important source of the perpetuation of both the bacteria is the infected or contaminated seed and infected plant residue. *X. axonopodis* pv. *phaseoli* (*Xap*) can survive in the infected seed for 30 years. As low as 0.5% infected seeds are sufficient to cause epiphytotic. Resident bacteria on weed hosts and non-host plants also act as a source of survival. When infected seeds germinate, the bacteria multiply and proceed as vascular pathogens throughout the large xylem vessels causing blocking of the vascular system and eventually wilting of the plants. In most cases, the vascular infection results from seed-borne inoculum, but it can also occur from penetration of bacteria through wounds. Stem cankers and leaf lesions may result from systemic infection. In humid weather, a yellowish bacterial ooze, that later dries to form a crust, is produced on the lesions of infected pods, leaves, stems, and cotyledons which serve as a source of secondary inoculum during the cropping season. The secondary spread of bacteria from these lesions occurs through disseminating agents. The dissemination of the bacteria occurs through the infected seed, wind-splashed rain, sprinkler irrigation, implements and insects. *Xap* can also be disseminated though the insects, such as *Diaprepes abbreviates* and *Cerotoma ruficornis*, grasshoppers (*Melanoplus* sp.), white flies (*Bemisia tabaci*), Mexican bean beetles (*Epilachna varivestris*), and leaf miners which may also provide routes for the pathogen entry by creating wounds while feeding. Seed is an important means for short, and long distance spread of the bacteria. The occurrence of devastating blight epiphytotic after wind-driven hailstorms is the result of the extensive spread of the bacteria by rain splashes and their entry through wounds caused by hailstorms. The penetration of bacteria into the host occurs through stomata, hydathodes, and wounds caused by insects, hailstorms, and cultural operations. The epidemiology of fuscous blight and common blight is similar. In general, the bacteria cause very severe disease under high rainfall and humidity and warm weather with optimum temperature (24°C–35°C) with maximum development occurring around 28°C.

Management

Cultural Methods

- Use disease–free seeds for planting. Do not use seeds from infected fields to plant in the next cropping.
- Select the fields in dry, aerated areas and with proper drainage.
- Rotate the legume crops with non-host crops for at least 2–5 years.
- Burn all debris of infected plants after bean harvesting. Destruction of infected plant debris can also be done by deep ploughing and burying it in the soil.
- Prune infected leaves to prevent the spread of disease within plants.
- Avoid cultural operations when the plants and fields are wet.
- Do not move farm tools through the infected fields.
- Remove and destroy the volunteer bean plants early in the following year and weed hosts.
- Escape the disease by choosing the suitable planting date.
- Avoid or limit the use of sprinkler irrigation.
- Do not use the run-off contaminated water for irrigation.

Host Resistance

- Grow resistant varieties.

Physical Control

- Seeds treatment with warm water is also a highly effective method in disease prevention. For this, dip the seed in hot water at 52°C for 20 minutes.
- Treat the seed with hot dry air at 60°C with 45%–55% RH for 23–32 minutes, which has been reported to be effective without affecting seed viability.

Chemical Control

The use of antibiotics is highly discouraged to avoid the development of antibiotic resistance in bacterial strains. Antibiotics should be used judiciously and in severe cases only.

- Treat seeds by dipping them in Streptomycin/Streptocycline solution (50,000 ppm) before planting because the seed dust may contain the bacteria.
- The Bronze mixture or Bronze sulphate is highly effective in preventing this disease and is used as a foliar spray on plants when early symptoms appear in the field.
- A preventive spray of Copper fungicides (Kocide, etc.) should be started early in the season and repeated at 7–10 days intervals to reduce the epiphytic population of the bacteria.
- Application of Bion (0.05%) and BioZell-2000B (50% etheric oil of thyme, 20% oil of maize, 20% oil of anise, 10% oil of sesame) is also effective in the reduction of bacterial blights.

20.2.2 Halo Blight

Halo blight of bean was first described by Burkholder (1926) from the United States. It is a major disease of beans throughout the world and causes serious economic losses (up to 43%) under favourable weather conditions. The occurrence of the disease has not been reported from India. Halo blight affects the foliage and pods of the common bean, lima bean, and soybean.

Symptoms

Leaf symptoms first appear as small, angular, water-soaked spots on the lower leaf surface. These lesions rapidly become necrotic, turn dry and reddish-brown, and are visible on both upper and lower leaf surfaces. Infection of expanding leaves may result in leaf distortion. A chlorotic zone of bright yellow tissue (halo) of irregular shape and size develops around necrotic reddish-brown lesions. Hence, the disease is called bacterial halo blight disease. The haloes, more prominent on young expanding leaves appear due to the production of a non-host selective toxin, phaseolotoxin by the bacterium and are a diagnostic symptom of the disease. At a temperature above 27°C, yellow halo forms very small or may be absent. In cases of severe infection, stabilised systemic chlorosis and death of new foliage can occur. Symptoms on pods and stems appear as water-soaked, red or brown, sunken lesions which may exhibit cream-coloured to silver, crusty bacterial ooze. Seeds in infected pods can be symptomless, wrinkled, discoloured (buttery-yellow patches on the seed coat), shriveled, and remains small. Seedlings raised from infected and or infested seeds, may show initial symptoms as water-soaked, round to irregular lesions on cotyledons and later general chlorosis, stunting, and distortion of growth.

Causal Organism and Etiology

Halo blight of beans is caused by *Pseudomons syringae* pv. *phaseolicola*. Taxonomy: Bacteria; Proteobacteria, gamma subdivision; order Pseudomonadales; family Pseudomonadaceae; genus Pseudomonas; species *Pseudomonas syringae*; Genomospecies 2; pathogenic variety phaseolicola. Microbiological properties: Gram-negative, non-sporing, strictly aerobic, motile, rod-shaped with rounded ends, 0.5–1.25 × 1.5–3.0 µm in diameter, at least one polar flagellum, optimal temperatures for growth of 25°C–30°C, oxidase negative, arginine dihydrolase negative, levan positive and elicits the hypersensitive response on tobacco.

Disease Cycle and Epidemiology

Sources of *Pseudomonas syringae* pv. *phaseolicola* are contaminated seeds and infected crop residue. The bacterium can survive for more than 4 years in infected seeds. One contaminated seed in 16,000 is sufficient to cause severe epiphytotic under favourable weather conditions. Infected plant debris on the soil surface also acts as a source for survival of the bacterium. Seedlings originating from contaminated seed harbor large numbers of bacteria. The bacterial ooze from seedling lesions serves as a source of secondary infections. The dissemination of the bacterium within the crop occurs through rain splash, wind, hail, blowing soil, farming tools, plant-to-plant contact, overhead irrigation and field workers. Bacteria enter the plant through natural openings or wounds during periods of high relative humidity or free moisture. Severe outbreaks of disease often occur after heavy rainstorms. The disease is favoured by cool to moderate temperatures (16°C–23°C) and humid moist conditions. Temperatures above 27°C inhibit the development of yellow haloes and systemic chlorosis. When the weather becomes warmer and drier, plants recover from the disease, and new growth is healthy. At cool temperatures, the pathogen produces a toxin, phaseolotoxin, which is responsible for systemic chlorosis. The production of phaseolotoxin is dependent on temperature, with 18°C–20°C being optimal and no detectable amounts of the toxin being present at 30°C. Phaseolotoxin inhibits ornithine carbamoyl transferase, a critical enzyme in the urea cycle.

Management

Management practices are similar to the common and fuscous blights of beans.

21 Ginger (*Zingiber officinale* Roscoe)

21.1 SOFT ROT

Soft rot is also called rhizome rot or *Pythium* soft rot of ginger. E.J. Butler (1907) recorded the incidence of this disease for the first time from Surat (Gujarat) in India. It is the most destructive disease of ginger, which can reduce production by 50%–100%. However, crop loss depends on the growth stage at which infection starts. Total loss results if the infection occurs in the early stage of crop growth. The loss can be as high as 90% during heavy infection. Under storage conditions, it can cause losses upto 50%–90%.

Symptoms

All the underground parts like roots, stems, and emerging sprouts are susceptible to this disease.

Symptoms appear initially as water-soaked patches at the collar region. These patches enlarge and the collar region becomes soft and watery, and then rots. Sprouts turn yellow and collapse. In mature plants, infection leads to the yellowing of leaves. This yellowing starts from the leaf tip and spreads downward, mainly along the margins resulting in the death of leaves. The dead leaves droop and hang down the pseudostem until the entire shoot becomes dry. Rhizomes first turn brown and gradually decompose, forming a watery mass of putrefying tissue enclosed by the tough skin of the rhizome (Figure 21.1). The fibrovascular strands are not affected and remain isolated within the decaying mass. Roots arising from the affected regions of the rhizome become soft and rot. The rotten parts emit a foul smell. Rotting attracts opportunistic fungi, bacteria, and insects. The basal portion of the plant exhibits a pale translucent colouration. This area later becomes water soaked and soft to such an extent that the whole shoot either topples or can easily be pulled out. Ginger plants are susceptible to *Pythium* infection at all stages of growth. If a diagnosis of the disease is just simply based on the above-ground symptoms, it could be mistaken for symptoms of either Fusarium yellows or bacterial wilt. See below the difference among these given in Table 21.1.

Figure 21.1 Soft Rot of Zinger.

Table 21.1 Difference in Pythium soft rot, Fusarium yellow, and bacterial wilt of zinger

Features	*Pythium* soft rot	Fusarium yellow	Bacterial wilt
Causal organism	*Pythium* spp.	*Fusarium oxysporum* f. sp. *zingiberi*	*Ralstonia solanacearum*
Symptoms	Yellow to golden yellow colour on the older leaves starting from the leaf margins, finally wilting and toppling of the plants apart from rotting of the rhizomes.	An initial yellowing at leaf margins on lower leaves with yellowing eventually extending to entire leaves and then wilting and drying of plants generally in patches within a field.	Infected plants show curling of the leaves, initially outwards, followed by yellowing and wilting.

Causal Organism and Etiology

About six *Pythium* species, namely, *P. aphanidermatum*, *P. butleri*, *P. delıense*, *P. myrlotylum*, *P. pleroticum*, *P. ultimum*, and *P. vexans* have been reported to cause soft rot in different parts of India. Among these species, the most common and destructive species in warm climates on ginger are *P. myriotylum* and *P. aphanidermatum*. *Pythium* is not a fungus; it is an oomycete (fungi-like organisms), related to algae. The common name is water mold.

Disease Cycle and Epidemiology

Pythium spp. are able to persist in soil, infected plant parts, and crop residue for long periods and the assumptions have been that this is mostly by means of encysted zoospores, oospores and sporangia, and serve as primary source inoculum. *Pythium* spp. can survive in air-dried soil for up to 12 years. The oospore germinates directly or indirectly. In the first case, the oospore produces a germ tube that elongates and either produces a sporangium or penetrates the host directly. In the second case, the oospore germinates, producing a sporangium and zoospores. The host root exudate causes the accumulation of zoospores around the root zone and induces encystment, germination, and infection. Young buds (or eyes), fine roots, developing rhizomes, and collar regions are the main points of infection. In addition, the appressoria are also produced by *Pythium* spp. for the infection process. The spread of disease is by waterborne zoospores or hyphal fragments. Oospore formation takes place in the host tissue or soil. When the seed rhizomes are infected, they fail to sprout due to the rotting of buds. After sprouting, the infection takes place through roots or collar region, finally reaching the rhizome. High soil water, high relative humidity, and relatively low temperature favour disease development and spread. Ginger planting often coincides with monsoon rains, and during this time, the soil water and ambient temperature (25°C to 30°C) become conducive to the onset of disease. Heavy rainfall and temperatures above 30°C (ideal conditions for rhizome rot development). Favorable conditions like wet soil conditions, high soil moisture, and soil temperature influence the development of the oospore. The severity of the disease is influenced by high rainfall and when rhizomes are planted in heavy clay soil with poor drainage. Temperature, 34°C is optimum for the germination of *P. aphanidermatum* and *P. myriotylum* spores.

Management

Cultural Methods

- Use of disease-free rhizomes for planting.
- Use raised beds.
- Do not plant, ginger crops in a field previously infested field with rhizome rot.
- Grow on zinger crops in well-drained soils and avoid water logging to prevent the development of rhizome rot and its spread. Proper drainage in sandy loam soil for cultivation ensures a healthy crop.
- Narrow ridge cultivation of zinger crops reduces the disease effectively compared to the unridged plots.
- Rotate the ginger crop with cassava, maize, cotton, yam, or legumes (these crops are not susceptible to rhizome rot) for at least 4 years. Avoid crop rotation with turmeric as this crop is also attacked by the disease.
- Use small, raised beds (30 cm in height, 1 m in width) with ditches around them.
- Avoid dense planting of the crop. Recommended spacing is at least 20–30 cm between plants.
- Keep fields weed-free.
- Cow dung slurry or liquid manure may be poured on the beds after each mulching to enhance microbial activity and nutrient availability.
- Maintain proper drainage.
- Adopt phytosanitary measures like infected plants should be uprooted and destroyed.
- Different types of cropping systems cropping like maize, papaya, cucumber, pumpkin, yam, tapioca, and different types of leguminous crops.
- Intercrop ginger with maize and pineapple.
- Use of resistant or tolerant varieties to rhizome wilt or rot.
- Ensure proper drainage. Adopt phytosanitary measures like infected plants should be uprooted and destroyed.

Physical Control

- Adopt soil solarization to reduce soilborne inoculum.
- Flooding treatment for 30 days, soil solarization during hottest months for 60 days.
- Treat the rhizomes with hot water at 47°C for 30 minutes.
- Treat the rhizome with hot water at 50°C for 10 minutes, at the time of sowing.

Host Resistance

- Grow resistant varieties to rhizome rot, if available.

Biological Control

- Use antagonistic fungi, namely *Trichoderma harzianum*, *T. hamatum*, *T. virens*, and bacterial isolates *Bacillus* and *Pseudomonas fluorescens* to suppress soil-borne pathogens of ginger.
- Apply *Trichoderma* sp. @ 2.5 kg/50 kg farmyard manure/ha to control *Pythium* and fertilize the soil or, apply neem-cake to the field at 200 kg/ha before 10–15 days of sowing.
- Application of *Boerhavia diffusa* leaves and seed powder of *Azadirachta indica* in the soil before planting reduces the infection.
- Adding neem cake to the soil also has been found to suppress *P. aphanidermatum*.
- Coating the seeds with *Trichoderma* spp. can reduce the soft rot.
- Seed (rhizome) disinfection with *Sodium hypochlorite* (1% solution) followed by soaking in *Trichoderma* spp. and followed by three applications of talc-based formulation (3×10^6 CFU/g) of *Trichoderma* to the soil at 15 days intervals from the time of planting reduces soft rot.
- Treating with *T. harzianum*, *Glomus mosseae,* and *fluorescent Pseudomonad* strain G4 together can inhibit the infection by upto 10%.
- Mulching with green leaves of *Vitex negundo* @ 4.0–4.8 t/acre is at the time of planting. (It is repeated @ 2 t/acre 40 and 90 days after planting).
- Bio-fumigation using cabbage and mustard plant refuses.
- Incorporation of neem cake and pine needle in the soil.
- Application of oil cakes made from *Azadirachta indica, Calophyllumino phyllum, Pongamia glabra, Hibiscus sabdariffa*, and *Brassica campestris.*
- Mulching with maha neem (*Melia azadirachta*) leaves (2.5 kg/m^2) can produce completely soft rot (*Pythium* spp.) free rhizomes.

Chemical Control

- Treat the rhizomes with Mancozeb or Carbendazim @ 0.3% for 30 min, prior to storing and planting. Carbendazim alone or in combination with Mancozeb can also be used.
- Soil drenching with Mancozeb (0.3%) or Cheshunt compound or Metalaxyl @ 500 ppm is effective in reducing disease incidence.
- Metalaxyl or Phosphorous acid in combination with Copper or bioagents has been used.

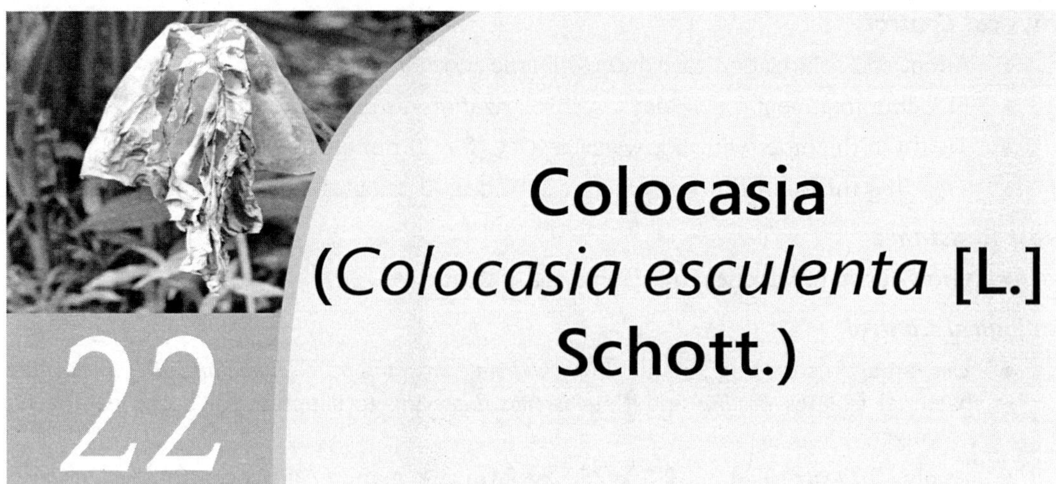

22 Colocasia (*Colocasia esculenta* [L.] Schott.)

22.1 PHYTOPHTHORA BLIGHT OR COLOCASIA LEAF BLIGHT

Leaf blight is the most destructive disease of taro causing a 25%–50% loss in yield and leaf yield upto 95% in susceptible varieties. E.J. Butler and G.S. Kulkarni (1913) reported leaf blight of taro for the first time in India.

Symptoms

The most obvious and frequent symptom is a blight of the leaf lamina, but *P. colocasiae* also produces a postharvest rot of the corms. The first symptoms appear as small, dark brown flecks or light brown spots on the upper leaf surface. These spots enlarge rapidly, becoming circular, zonate, and purplish brown to brown, often with yellowish margins. On the lower surface of a leaf, the spots have a water-soaked or dry grey appearance and sometimes contain hard globules of plant exudates. As the disease progresses, the lesions (mostly at the tip and leaf margin) continue to expand and frequently coalesce and quickly destroy the leaf. Diseased tissues become papery, disintegrate and drop out, forming "shot holes" of irregular size and shape on the affected leaves (Figure 22.1). Dead leaves often hang on their long petioles like flags. Another conspicuous and characteristic feature of Phytophthora blight of colocasia is the formation of amber, bright orange, or reddish brown plant exudate oozing from infected tissues. These droplets dry out during the day to form crusty deposits on the surface of the lesion. A whitish ring of mycelia and sporangia which is a prominent sign of *P. colocasiae,* around the border of lesions appears under high humid and wet conditions. Occasionally the pathogen may cause water-soaked lesions on the petioles. Grey-brown to dark-blue lesions occurs on undamaged harvested corms. These lesions enlarge rapidly and coalesce. Affected corms are almost completely decayed in wet and warm conditions in 7 to 10 days. In the early stages, the diseased tissue is light-brown, firm, and often has a distinct margin.

Figure 22.1 Phytophthora Blight of Colocasia or Taro. (A) Typical symptoms of Phytophthora Blight on leaf, (B) Severely blighted colocasia leaf.

Causal Organism and Etiology

Taro leaf blight (TLB) is caused by the fungus-like organism *Phytophthora colocasiae* (An Oomycetes). The mycelium is hyaline, coenocytic, and inter- or intra-cellular. The haustoria are slender, long, and unbranched. The sporangiophores are very slender, unbranched, and extremely narrow. The sporangia are ovoid to ellipsoid with a distinct narrow apical plug (semi-papillate). Chlamydospores are thick-walled, round, and hyaline. The oogonium is spherical and yellowish and the amphigynous antheridium persists at the base of the oogonium.

Disease Cycle and Epidemiology

The primary mode of survival is the continuous recycling of the pathogen, often on single plants within the crop. However, the pathogen also survives in infected crop residues left in the field, rhizomes, and soil, as mycelium, encysted zoospores, chlamydospores, and oospores. *P. colocasiae* also survive saprophytically and on the alternate host. The primary reproductive unit of the pathogen is the sporangium which requires free water to germinate. The waxy surface of taro leaves accumulate tiny droplets of water that provide sufficient moisture for the germination of sporangia and movement of zoospores. Sporangia can germinate directly or indirectly depending on the weather conditions. Under cooler conditions (20°C–21°C), the sporangia germinate indirectly by producing remiform, biflagellated, 15–20 zoospores. These lose their flagella and form a rounded cyst which soon germinates to form a germ tube and cause infection. Under warmer conditions (about 25°C), the sporangia germinate directly by germ tubes that can infect the leaves. Sporangia and zoospores are spread by rain splash and wind-blown rain between plants or on the same plant. Long-distance dispersal of the organism occurs only by the movement of infected plant material (leaves, petioles, infected corms). Long periods of leaf wetness, rainfall, high humidity (close to 100%) and temperature (20°C–30°C) are the key factors controlling the disease cycle and epidemiology of *P. colocasiae*.

Management

Cultural Methods

- Use of disease-free planting material.
- Remove crop debris left in the field from the previous crop.
- Remove all or parts of infected leaves to reduce inoculum levels. As in the early season, sporulating leaf lesions supply enough propagules (sporangia, zoospores) to increase disease.
- Early planting to avoid heavy monsoon rains.
- Remove self-grown colocasia plants.
- Avoid excessive levels of moisture in the field.
- Hot water disinfestations.
- Follow crop fallows for disease suppression.
- Avoid dense planting of taro.
- Isolate plantings (e.g., three small, separate patches instead of one large patch).
- Prepare the soil well and amend it before planting if calcium, magnesium, or phosphorus is needed.
- Monitor plant calcium levels by leaf analysis, and maintain calcium at recommended concentrations to prevent the development of *Pythium* corm rot.
- Add lime material before planting to raise soil pH to 6.0–6.8.
- Rotate taro with other crops.
- Incorporate compost and apply surface mulch. Mulching with paddy straw slightly lowers disease severity.
- Rogue diseased plants, take them far from the planting site and destroy them by burying or burning (if allowed), or composting.

Biological Control

- Treat the seed (rhizome) with biocontrol agents viz., *Trichoderma viride*, etc.

Host Resistance

- Grow resistant or tolerant varieties such as Jhankri, Sonajuli, Muktakeshi, Satamukhi, Bhavapuri (KCS-2), Indira Arvi-1, Narendra Arvi-1, Narendra Arvi-2, Upland Taro-Bidhan Chaitanya, and Upland Taro.

Chemical Control

- Most commonly recommended fungicides are Mancozeb (e.g. Dithane M45), Copper (e.g. Copper oxychloride), Metalaxyl (e.g. Ridomil MZ–containing Mancozeb), and Phosphorus acid (e.g. Phoschek).
- Spray Blitox-50 or Fytolan @ 3 g/L water or Dithane Z-78 @ 2 g/L water at 10–15 days intervals. Add stickers like Rosin @ 4 g/L of solution or Sadovit @ 1 ml in 2 litres of solution.
- Spray with fungicides viz., Mancozeb (0.2%) or Ridomil MZ 72 @ 2 g/L of water.

23 Coconut (*Cocos nucifera* L.)

23.1 WILT: TANJORE WILT OR GANODERMA WILT OR BASAL STEM ROT

Coconut wilt is also known as Tanjore wilt, or Basal stem end rot, or Ganoderma wilt. Basal stem rot (BSR) or wilt or root rot is the most destructive disease of the coconut palm. *Ganoderma* spp. has a wide host range attacking a variety of palms and several forest, avenue, and fruit trees. In India, *Ganoderma lucidum* on coconut was first recorded in Karnataka state during the year 1913 by Butler. Ganoderma wilt of coconut was first observed in the coastal area of Tanjore/Thanjavur district of Tamil Nadu in 1952 and due to this reason, it was named as Thanjavur wilt. The disease is also known as Ganoderma wilt (Andhra Pradesh) or Tanjore/Tanjavur wilt (Tamil Nadu) or bole rot or anabe roga (Karnataka) in different parts of India. BSR of coconut is a serious disease in India, and in severely infected areas, the incidence was recorded as high as 80% in 1991.

Symptoms

The most characteristic initial symptom is withering followed by yellowing and drooping of leaves in the outer whorl. Even before this, the decay of finer roots would have started. The root decay gradually extends up to the bole region. The yellowing and withering extend to the inner whorl of leaves also. This is accompanied by the exudation of reddish-brown viscous fluid from the basal portion of the trunk. The decay of the bole slowly goes upwards into the basal portion of the stem. Some palms show wilting symptoms without any external signs of bleeding, but with extensive internal decay. Palms exhibiting bleeding symptoms without any drooping of leaves are also very common. The bleeding patch enlarges in size and spread upwards, to a height of 4 meters in certain cases. There is a commensurate internal decay just below the bleeding patch. Severe drooping of leaves often results in nut fall. The spindle size is reduced and so is the crown size also. The bark turns brittle and often gets peeled off in flakes, leaving open cracks and crevices. The internal tissues are discoloured, disintegrated, and emitting a bad smell. Bracket formation occurs at the base of the trunk during the rainy season. Ultimately the palm dies off. The exposed tender tissues in the crown are easily attacked by soft rotting bacteria/fungi and this results in the crown emitting a foul smell. Due to this weakening, the crown topples over in the slight wind. In a few diseased palms in the advanced phase, or dead palms, brackets of *Ganoderma lucidum* can be observed at the base.

Causal Organism and Etiology

Basal stem end rot of coconut is caused by *Ganoderma lucidem* and *Ganoderma applanatum.*

The aerial mycelium of *Ganoderma* is hyaline, thin-walled, branched with frequent clamp connections, abundantly formed chlamydospores which are slightly thick-walled, terminal or intercalary, ellipsoid and sometimes in chains. The fruiting body is perennial, sessile, reniform to irregular, dimidiate basidiocarps stipitate, usually lateral, corky becoming woody later, usually 10–12 × 3–4 cm, but may grow up to 30 cm or more, the upper surface is glossy, laccate crust, and reddish brown, but darkens with age. Basidiospores are pale yellow, ellipsoid, with long and thick echinules.

Disease Cycle and Epidemiology

Ganoderma spp. produces various resistant stages such as resistant mycelium, chlamydospores, basidiospores and pseudosclerotia, and survives in infected plant parts, crop debris, fruiting bodies (basidiocarps), and soil. The basidiospores of *Ganoderma* spp. probably play a major role in the dispersal and initiate the infection process. Basidiospores after germination cause infection through wounds on roots, trunks, or cut ends of palm fronds. Temperature, rainfall, rainy days, and relative humidity, play a significant role in disease development. Soils with poor drainage and water stagnation during rainy seasons were found to favour the disease. Symptoms of BSR disease can take several years to develop, and the presence of the pathogen (such as indicated by fruiting bodies) is often only visible when the fungus is well established and more than half of the bole tissue has been decayed, leaving no chance for the grower to cure the infected palms. It spreads mainly through root contact and irrigation between diseased and healthy palms. Repeated ploughing and uncontrolled flood irrigation, neutral pH, sandy loam soil, and high plant density in plantation also favours the disease.

Management

Cultural Methods

- Remove and destroy all affected palms.
- Avoid ploughing and flood irrigation.
- Avoid flood irrigation and repeated ploughing.
- Digging isolation trenches around the infected tree is recommended. Isolation trenches (1 cm × 30 cm) may be dug up around diseased palms to prevent root contact.

- Intercrop with banana. Growing intercrops such as *Desmodium tortuosum, Calopogonium mucunoides, Tephrosia purpurea, Crotalaria juncea, Curcuma domestica* [*Curcuma longa*], and *Musa* sp. reduced disease incidence. Among the intercrops, the banana was very effective in reducing the disease.
- Green manure crops (sunhemp and sesbania) must be raised and ploughed in situ before flowering.
- Application of 50 kg farmyard manure or green leaves or 5 kg neem cake or 300 kg tank silt per palm per year arrested disease progress.
- Apply 2 kg of lime and 3.5 kg Potash/palm.

Biological Control

- Apply *Pseudomonas fluorescens* (Pf-1) @ 200 g per palm + *Trichoderma viride* @ 200 g per palm/year.
- Apply 200 g phosphobacteria and 200 g *Azotobactor* mixed with 50 kg of FYM per palm.
- Apply FYM 50 kg + neem cake 5 kg once in 6 months along with fertilizers.
- Apply *Pseudomonas flourescens* (PF1) @ 100 g per palm/year + *Trichoderma viride* @ 100 g per palm/year in soil.

Chemical Control

- The bleeding patches in the stem may be chiseled and protected with Tridemorph (5% Calxin) and subsequently with hot coal tar.
- Apply Aureofungin-sol 2 g + 1 g Copper Sulphate in 100 ml of water/tree or Tridemorph 20% EC @ 10 ml + 15 ml of water/tree as root feeding. (The active absorbing root of pencil thickness must be selected and a slanting cut is made. Keep the solution in a polythene bag or bottle and the cut end of the root should be dipped in the solution).
- Root feeding with Tridemorph 2 ml or Hexaconazole 1 ml with 100 ml of water (3 times at 3 months intervals).
- Forty litres of 1% Bordeaux mixture should be applied as a soil drench around the trunk in a radius of 1.5 meters.
- Bordeaux mixture (1%) and Aureofungin-sol (0.2%) chemicals alone or in combination are effective chemicals.
- Drench the soil with 10 litres of Benomyl (0.1%).
- Aureofungin-sol (2 g) + 1% Bordeaux mixture (40 litres) + neem cake (5 kg) check further the spread of disease.

23.2 BUD ROT

In India, bud rot of coconuts was reported by E.J. Butler in 1906. Bud rot is distributed worldwide, and severe infestations can cause total loss of a crop. It is generally a sporadic disease but sometimes appears in epidemic forms. The disease commonly occurs on West and

East Coasts of India. Bud rot disease incidence in coconut recorded upto 15%–20% in Andhra Pradesh.

Symptoms

First symptom of bud rot disease is yellowing and withering of one or two young leaves in the spindle. Later, the lesion of the spear leaf or spindle turns brown/black followed by bending and drooping of spindle. The basal tissues of the leaf rot quickly and can be easily separated from the crown. The spindle withers and droops down and one by one, the inner leaves also fall away, leaving only fully matured leaves in the crown. A foul smell is emitted by the internal rotted tissues of the affected areas. Ultimately the palm succumbs to the disease with the death of the growing bud. As the disease progress, the infection extends to the older leaves, causing sunken leaf spots covering the entire leaf blade and spreading both up and down. The spot margins are irregular and water soaked and the spots become conspicuous on the blade of unfolded leaves. In severely affected trees, the entire crown may rot, and trees wilt. The heart leaf becomes chlorotic, wilts, and collapses. Light brown to yellow, oily, sunken lesions may be found on leaf bases, stipules or pinnae. The pathogen also causes bud rot and premature nut fall during the rainy season. Water-soaked, greyish-green lesions develop at the stalk end of the nuts which leads to premature nutfall. Later these lesions turn brown and become sunken due to the decay of underlying tissues. The rot extends into the husk and sometimes deep into the endosperm cavity.

Causal Organism and Etiology

Bud rot of coconut is caused by *Phytophthora palmivora*. It produces sporangia: non-caducous, papillate, ovoid and bilaterally symmetric. It is homothallic and produces oogonia with varying numbers of irregular surface protuberances and tapered, funnel-shaped bases. The antheridia are amphigynous, and oospores are aplerotic.

Disease Cycle and Epidemiology

The lifecycle of *P. palmivora* in the tropics is complex due to a large host range and multiple tissues of the same host plant affected, such as roots, stems, leaves, flowers, and fruits. The pathogen survives in the disease-affected crown tissues during the summer months and multiplies during the ensuing season. The pathogen also survives in other hosts as it has a wide host range such as oil palm, arecanut, palmyra, citrus, rubber, cocoa, black pepper, bougainvillea, cardamom, hibiscus, jackfruit, etc. The pathogen survives in the form of oospores and chlamydospores in soil and coconut debris such as cull piles, and infected roots, serve as reservoirs of primary inoculum. These germinate producing sporangia and later zoospores which are spread by wind, wind-driven rain, and possibly by insects from the family Tettigonidae. Insects may be involved in the dispersal of propagules of *P. palmivora* from the soil to the top of the palms, where the pathogen can infect the young spear leaf tissue. As the zoospores reach the trichomes they germinate and form an appressorium-like structure that subsequently penetrated the tissue through stomata. After penetration, hyphae colonized the intercellular leaf spaces, followed by intracellular invasion, penetration of the vascular bundles, and spread throughout the host tissue. Sporangia, chlamydospore, and zoospore start a cycle of secondary infections. Secondary infections, typically below the initial point of infection, develop several weeks later under regular misting conditions that promote sporangia production and wash down to the base of newly forming leaflets. A number of different spores such as zoospores, sporangiospores, chlamydospores, and oospores can be produced, and airborne, soilborne, water-borne, and vector-borne spread of propagules is possible. Secondary infections are repeated several times under favourable environmental conditions, resulting in small clusters of lesions along the length of the emerging spear leaf. Disease development coincides with the prevalence of high humidity and low temperature. Young palms are more susceptible to this disease. Heavy rainfall, high relative humidity (94%–100%) and temperatures below 24°C are highly favourable for disease development and spread.

Management

Cultural Methods

- Adopt proper spacing and avoid over-crowding in bud rot prone gardens.
- Adopt strict phytosanitary measures for the effective prevention of the disease.
- Remove and burn palms killed due to bud rot along with decayed tissues.
- Apply salt and ash mixture or paddy husk after removing the affected portion in the crown and subsequently cover with a mud pot. This will absorb moisture and keep the protected portion dry.
- Remove all the dead palms and palms beyond recovery and dispose them by burying or burning them.
- Clean the crowns of the palm of any debris and other materials.
- Maintain general field hygiene by removing the weeds etc.
- Follow recommended integrated nutrient management (INM) practices.
- Provide adequate drainage in case of low-lying areas.
- Drench the crown with 2% *Pseudomonas fluorescens* or PGPR Mix 2 as a prophylactic measure before the onset of monsoon in endemic areas.
- Adopt control measures for the Rhinoceros beetle which act as a predisposing factor for infection.

Chemical Control

- Spray Copper oxychloride 50% WP @ 1 kg in 300–400 litres of water/acre on the crown of the neighbouring palms as a prophylactic measure before the onset of monsoon.
- Remove all the affected tissue of the crown region and drench the crown with Copper oxychloride (0.25%).
- Apply Bordeaux paste and protect it from rain till a normal shoot emerges. (Dissolve 100 g of Copper sulphate and 100 g of quick lime each in 500 ml water separately and mix to form 1 litre of Bordeaux paste).
- Spray 0.25% Copper oxychloride or 1% Bordeaux mixture on the crown of the neighboring palms as a prophylactic measure before the onset of the monsoon. Palms that are sensitive (Dwarf palms) to copper containing fungicides can be protected by Mancozeb. Small, perforated sachets containing 2 g of Mancozeb may be tied to the

top of leaf axil. When it rains, a small quantity of the fungicide is released from the sachets to the leaf base, thus protecting the palm.

- The infected tissues from the crown region should be removed and dressed with Bordeaux paste sprayed with 1% Bordeaux mixture as pre-monsoon spray (May and September).
- Leaf axil filling with Sevidol 8G, 25 g mixed with 200 g sand is recommended to manage red palm weevil infestation of affected palms.
- Spray Copper oxychloride (0.25%) after the onset of the monsoon.
- The infected tissues in the crown are to be removed and the wound dressed with 10% Bordeaux paste and covered with perforated polythene sheet till the next leaf emerges.
- Prophylactic spraying with 1% Bordeaux mixture should be carried out to all the trees around the infected palm.
- Apply bleaching powder on the affected portion after removing the infected tissues.
- Place perforated polythene sachets containing an aqueous paste of Mancozeb (@ 2 g per sachets) in the 2–5 top most leaf axils of palms with the onset of monsoon.
- As a prophylactic measure, adjacent healthy palms should be sprayed with 1% Bordeaux mixture or with any other copper-based fungicide like Copper oxychloride @ 2 g/L of water. A pre- and post-monsoon sprays of the above fungicide are recommended for the management of the disease.
- Spray palms with Dimethomorph @ 1 g/L of water or Metalaxyl 35 EC @ 2 g/L of water.
- Manage the rhinocerous beetle in the FYM/compost yard by regularly treating the compost pit with bioagent (*Metarhizium anisopliae*) or drenching with Chlorpyriphos (0.05%). This will reduce the population of the insect pest in the vicinity, so that the damage caused by the adult beetle in the coconut plantation is reduced.

24 Tea (*Camellia Sinensis* [L.] O. Kuntze)

24.1 BLISTER BLIGHT

Blister blight is a leaf disease in tea that preferentially attacks the economically important young leaves and is by far the most serious disease of cultivated tea. If not controlled by fungicides, tea losses due to blister blight may be as high as 35%. The most common leaf disease blister blight caused by *Exobasidium vexans*, by far the most important leaf disease, is known to occur in almost all the tea growing areas in India, Sri Lanka, and Japan. Blister blight is capable of causing enormous crop loss throughout the tea growing regions of Asia, especially in India, Sri Lanka, Indonesia, and Japan. Since the pathogen attacks harvestable tender shoots, it inflicts a global yield loss of 40%. Symptoms of blister blight were described in detail for the first time by Petch (1923). The disease was first reported in Assam in 1868. Sir George Watt was the first to report the disease symptoms in Assam in the year 1895. In 1908, a sudden outbreak of the disease occurred in Darjeeling, West Bengal, India.

Symptoms

The young and tender shoots are most susceptible and attacked by the pathogen. The initial symptoms of blister blight appear as a small, pale-green, pale-yellow, or pinkish, translucent spot on the tea leaf. These tiny pinhead-sized spots are referred to as the first stage of the disease. As the leaves develop, the spots become transparent, larger (upto 3.0–12.5 mm in diameter), and light brown. Later, blister-like symptoms, with dark green, water-soaked zones surrounding the blisters, develop on the lower surface of the leaf. Following the release of the fungal spores, the blister becomes white and velvety. Subsequently, the blister turns brown. On the stem initially, a pale-yellow spot appears, these spots gradually elongate and encircle the whole stem, become slightly swollen and ultimately turn grey due to the maturation of spores.

Finally, the stem bends over, and breaks off or dies; consequently, the death of young shoots occurs and results in more serious crop loss.

Causal Organism and Etiology

Tea blister blight is caused by *Exobasidium vexans.* It is an obligate parasitic fungus systematically placed under Exobasidiaceae, Exobasidiales, Exobasidiomycetes, and Basidiomycota. Hymenium develops under the epidermis of stem or lower side of leaf, erumpent. Paraphyses single, septate, apically rounded, basidia clavate, generally bearing two sterigmata, projecting above paraphyses, basidiospores ellipsoid, initially, hyaline, unicellular, becoming single septate at maturity. A transverse section through the blister region of the leaf showed that mycelium grows intercellularly and produces finger-like haustoria which penetrate the leaf parenchyma cells. The hymenium forms below the epidermis on the lower surface of the infected leaf. A palisade of paraphyses and basidia arises, forcing up the epidermis which forms the blister and then ruptures. Paraphyses are single, septate, and apically rounded. The basidia are clavate, and generally bear two sterigmata. Basidiospores are unicellular at first, but a septum develops on maturity; they are ellipsoid, initially hyaline.

Disease Cycle and Epidemiology

Blister blight is a polycyclic disease with a relatively short (11–28 days) fungal life cycle. *Exobasidium vexans* completes its life cycle on the tea plant only, because it is an obligate parasite with no known other host (alternate or collateral host). The fungus perennates in infected branches and necrotic blister lesions of the tea bushes as mycelia or thick-walled spores. During favourable climatic conditions, the perennating mycelium becomes active and grows intercellularly until the development of a palisade of paraphyses and basidia below the epidermis of the lower surface of the tender leaves. During this phase fungal mycelia grow rapidly within the host tissue; host tissue develops a blister-like structure just below the lower epidermis. Then, the epidermis breaks to expose the clavate basidia and basidiospores. In addition to basidia, the hymenial layer possesses conidiophores bearing two-celled conidia. The mature blister lesion can produce two million spores in 24 hours. Large masses of basidiospores spread rapidly by wind and cause secondary infections and repeat continuously several disease cycles, under favourable (wet) conditions. The wind-borne spores land on a susceptible host tissue with adequate moisture; germinate and infect and produce visible symptoms within 10 days. The infection is facilitated by the formation of an infection peg from appressoria either directly penetrating the cuticle of host tissue or penetrating through stomata. After penetration visible sign of infection appears as translucent spots, resulting in the development of characteristic blisters on the lower-surface of the young leaves. At the cellular level, the enlargement of the infected spots caused by an infection of *Exobasidium* species is either due to hypertrophy or hyperplasia–an increase in the number of cells. Owing to the short span of the life cycle, multiple generations of the pathogen are completed within a single cropping season. Monsoon is the favourable time for infection, sporulation and spore dispersal. Low temperature (20°C–25°C), high humidity (more than 80%), cloudy weather conditions with moderate rainfall, play a profound impact on the development of pathogens, and disease incidence.

Management

Cultural Methods

- Use healthy planting material.
- Adopt early pruning and hand plucking to reduce the disease severity as the pathogen infects only tender shoots.
- Collect and destroy all the blister infected leaves.
- Prune immediately severely infected tender young tea plants. Pruning during November/ December is effective to reduce the disease incidence for new clearing.
- Avoid broad-leaved Assam jats.
- Prohibit the entry of workers from the infested section into the healthy sections.

Biological Control

- Spray 2–3 rounds of 5%–10% aqueous extracts of *Cassia alata* or *Polygonum hamiltoni* or *Acorus calamus* or *Adhatoda vasica* or *Equisetum arvense* or *Polygonum hydropiper* or *Tagetis petula* at 15 days interval.
- Antagonists like *Trichoderma harzianum*, *Gliocladium virens*, *Serratia marcescens*, *Pseudomonas fluorescens*, and *Bacillus subtilis* have been used experimentally for managing blister blight of tea.
- Use of *Pseudomonas fluorescens* and *Bacillus subtilis* as liquid culture supplemented with Ammonium sulphate and salicylic acid has been shown to enhanced bio-efficacy of bioagents and reduced disease incidence.
- A phylloplane bacterium *Orchobacterium anthropi* BMO-111 has been shown to achieve better performance compared to conventional chemical fungicides Copper oxychloride and Hexaconazole.

Chemical Control

- Spray Bordeaux mixture or Copper oxychloride @ 0.1%
- Spray a mixture of Copper oxychloride (210 g) + Nickel chloride (210 g) per hectare at 5 days intervals from June to September and October to November.
- Spray of Tridemorph @ 340 ml/ha and 560 ml/ha is satisfactory under mild and moderate rainfall conditions.
- Bitertanol 25% WP @ 80 g in 30 litres of water/acre or Copper oxychloride 50% WP @ 0.168 g in 70 litres of water/acre or Copper hydroxide 77% WP @ 140 g in 300 litres of water/acre or Hexaconazole 5% EC @ 80 ml, 28–36 with power sprayers 70–80 with knap-sack sprayer/acre or Propiconazole 25% EC @ 50–100 ml in 70–100 litres of water/acre.
- Spray Copper oxychloride (50% WP) @ 350–420 g in 67 litres of water per hectare with an air blast sprayer, covering two rows on either side.

25 Coffee (*Coffea canephora* L. and *Coffea arabica* L.)

25.1 LEAF RUST

It was first discovered in the vicinity of Lake Victoria in East Africa in 1861 and later identified and studied in Sri Lanka (previously known as Ceylon) in 1867. In 1869, the first coffee leaf rust outbreak was reported on the Asian continent, specifically in Sri Lanka. Coffee leaf rust was introduced in India from Sri Lanka in 1879. The two main cultivated coffee species, *Coffea canephora* (*Robusta coffee*) and *C. arabica*, account, on average, for 40% and 60% losses, respectively, of the world's coffee production.

Symptoms

The first observable symptoms are small (2–3 mm diameter), pale yellow spots on the upper surfaces of the leaves. Later these spots gradually increase (upto 15 mm diameter), and powdery masses of orange-yellow to red-orange urediniospores (uredospores) appear on the undersurfaces (Figure 25.1). Blotches may merge and cover the leaf blade. The fungus does not produce typical pustules as many rusts, because it sporulates through the stomata rather than breaking the epidermis as most rusts do. In severe infection, barriers remain small and fail to ripen, leaves fall, severe twigs dieback and the death of the tree occurs.

Causal Organism and Etiology

Coffee leaf rust is caused by *Hemileia vastatrix*. The genus *Hemileia* is a member of the phylum Basidiomycota, class Pucciniomycetes, order Pucciniales (rust fungi). *Hemileia vastatrix* is is a biotrophic and microcyclic (pycnial/spermagonial and aecial stages absent) pathogen. Urediniospores or uredospores: produced in clusters and emerge through stomata; not typically round to oval like other rust fungi, but more or less reniform (kidney-shaped) with a curved surface covered in short spines and a smooth, flattened surface–hence the generic name *Hemileia*, meaning 'half smooth'. Teliospores: pedicellate, subspherical or napiform (turnip-shaped), smooth-walled. Teleiopsores germinate and produce basidia on which 4 basidiospores form. Urediniospores and teliospores are produced in the same sorus but at different times.

Figure 25.1 Coffee Leaf Rust. (A) Powdery yellow-orange pustules on lower surface of coffee leaf, (B) close-up of pustules. *Courtesy:* Dr. Elijah Gichuru, Coffee Research Institute, Kenya Agricultural and Livestock Research Organization, Ruiru, Kenya.

Disease Cycle and Epidemiology

Hemileia vastatrix is a microcyclic/hemicyclic fungus producing urediniospores, teliospores and basidiospores, whereas pycniospores and aeciospores are not known. The coffee rust fungus exists primarily as mycelium, uredia, and urediospores on infected leaves. The *Hemileia* fungus cycle begins with the process of releasing and landing a spore on the coffee leaf; subsequently, the spore germinates and the infection process begins. Urediniospores are dikaryotic and represent the asexual cycle, re-infecting the leaves whenever environmental conditions are favourable. Basidiospores cannot infect coffee, but no other host plant has been identified. The initiation of *H. vastatrix* infection on coffee leaves, like other rust fungi, involves specific events, including adhesion of urediniospore to the host surface, germination, appressorium formation, penetration, and inter- and intra-cellular colonization. Uredospores germinate in presence of free water and produce germ tubes. The germ tubes, either directly or after branching, produce appressoria over stomata. After appressorium formation, the fungus penetrates the host through the stomata, forming an infection hypha that grows into the substomatal chamber. Infection hypha produces two thick lateral branches, resembling an anchor, a unique trait of *H. vastatrix*. Each lateral branch of the anchor differentiates into a haustorial mother cell that gives rise to a haustorium, which primarily infects the stomatal subsidiary cells, another unique feature of *H. vastatrix*. The fungus continues to grow, forming more intercellular hyphae and a large number of haustoria in the cells of the spongy and palisade parenchyma and even of the upper epidermis. This leads to the appearance of a lesion on the leaf surface. Urediniospores initiate infections that develop into lesions that produce more urediniospores. It takes 10 to 14 days from infection for new uredinia to develop and urediniospores to be formed. The rust lesions continue to enlarge over a period of 2 to 3 weeks. A single lesion can produce 4–6 crops of spores, releasing about 300,000 urediniospores over a period of 3–5 months. Secondary cycles of infection occur continuously during favourable weather, and the potential for explosive epidemics is enormous. The uredospores are locally disseminated by rain splashes, thrips, flies, wasps, and other insects and over long distances (intercontinental, across oceans, deserts, and mountain ranges) by wind and human intervention. Temperature also influences the normal urediniospore germination and other infection processes greatly which only occur between 15°C–30°C. Urediniospores germinate only in the presence of free water (rain or heavy dew); high humidity alone is not enough. Continuous free moisture is required for 24–48 hours to complete the process of infection. Coffee leaf rust infections seldom kill the host plant, although severe infections affect vegetative development and can generate polyetic epidemics over successive seasons.

Management

Cultural Methods

- Sanitation of the plantation is essential.
- Collect infected fallen leaves and destroy them.
- Wider spacing and appropriate pruning.

Host Resistance

- Grow resistant varieties such as S 238, S 395, etc.

Biological Control

- Hyperparasitic fungi, *Verticillium hemileiae*, *V. lecani*, *V. leptobactrum*, *V. psalliotae*, *Cladosporium hemileiae*, *Paranectris hemileiae*, and *Darluca filum* possibly can be used as bioagents against rust fungi. With coffee rust, hyperparasitism of *Verticillium hemileiae* reduces the viability, penetrates the uredospores and kills it, but it has very little impact on overall rust development.
- *Bacillus subtilis* and *Bacillus megaterium* are pathogens of rust fungus and may be used as bioagents.
- Coffee leaf extract, *Bacillus subtilis* and *Pseudomonas putida* induce host resistance, and reduce diseases.

Chemical Control

- Spray Copper oxychloride 50% WP @ 1.5–2.2 kg in 300–400 litres of water/acre.
- Spray Bordeaux mixture @ 0.5% before flowering, during rainy and after rainy seasons in May, August–September and October respectively to prevent the diseases.
- Spray Carboxin (Plantvex) 0.01% a.i. or Oxycarboxin (Vitavex) @ 0.03% a.i.
- Spray of host resistance activator Acibenzolar-S-methyl has been reported to induce host resistance against coffee leaf rust.

26 Lists of Important Diseases of Crops

26.1 RICE (*ORYZA SATIVA* L.)

Disease		Causal Organism / Agent
Fungi		
Blast	:	*Pyricularia oryzae* (*syn. Magnaporthe oryzae*)
Brown spot or Sasame leaf blight or Nai-Yake	:	*Bipolaris oryzae* (syns. *Cochliobolus miyabeanus, Helminthosporium oryzae*)
Sheath blight	:	*Rhizoctonia solani* (syn. *Thanatephorus cucumeris*)
False smut	:	*Ustilaginoidea virens* (teleomorph form: *Villosiclava virens*)
Bakanae	:	*Fusarium fujikuroi* (syn. *Gibberella fujikuroi*)
Sheath Rot	:	*Sarocladium oryzae*
Leaf smut	:	*Eballistra oryzae* (syn. *Entyloma oryzae*)
Stackburn	:	*Trichoconiella padwickii* (syns. *Alternaria padwickii*)
Stem rot	:	*Sclerotium oryzae*
Scab/head blight	:	*Fusarium graminearum*
Aggregate sheath spot (Brown sclerotial disease)	:	*Ceratobasidium setariae* (syns. *Ceratobasidium oryzae-sativae; Rhizoctonia oryzae-sativae*)
Kernel smut	:	*Tilletia barclayana* (syns. *Neovossia barclayana*), *Tilletia horrida* (syn. *Neovossia horrida*)
Sheath spot (Bordered sheath spot, red sclerotial disease, Rhizoctonia sheath spot)	:	*Waitea circinata* (syn. *Rhizoctonia oryzae*)

Disease	Causal Organism / Agent
Udbatta (Black choke, incense rod, false ergot)	: *Balansia oryzae* (syns. *Balansia oryzae-sativae*; *Ephelis oryzae*; *Ephelis allid*)
Narrow brown leaf spot	: *Cercospora janseana* (syns. *Cercospora oryzae*)
Bacteria	
Bacterial leaf blight (BLB)	: *Xanthomonas oryzae* pv. *oryzae*
Bacterial leaf streak (BLS)	: *Xanthomonas oryzae* pv. *oryzicola*
Bacterial panicle blight and grain rot	: *Burkholderia gladioli*, *Burkholderia glumae*
Virus	
Rice tungro disease (dual infection)	: Rice tungro virus (RTSV, RTBV) *Rice tungro bacilliform virus* (RTBV)—genus *Tungrovirus*; family *Caulimoviridae* *Rice tungro spherical virus* (RTSV)—genus *Waikavirus*; family *Secoviridae*
Dwarf	: *Rice dwarf virus* (RDV)—genus *Phytoreovirus*; family *Reoviridae*
Grassy stunt	: *Rice grassy stunt virus (RGSV)*—genus Tenuivirus; family Phenuiviridae
Hoja blanca	: *Rice hoja blanca virus (RHBV)*—genus Tenuivirus; family Phenuiviridae
Nematode	
Root-knot nematodes	: *Meloidogyne* spp. (*Meloidogyne arenaria*, *M. graminicola*, *M. incognita*, *M. javanica*, *M oryzae*, *M salasi*
White-tip nematode	: *Aphelenchoides besseyi*
Rice root nematode	: *Hirschmanniella oryzae*
Disorder/deficiency	
Khaira	: Zinc deficiency

26.2 MAIZE (*ZEA MAYS* L.)

Disease	Causal Organism / Agent
Fungi and Oomycetes	
Pythium stalk rot	: *Pythium* spp. (*Pythium aphanidermatum*)
Fusarium and Gibberella stalk rots	: *Fusarium moniliforme* syn. *Fusarium verticillioides* (Teleomorph: *Gibberella fujikuroi*), *Fusarium graminearum* (Teleomorph: *Gibberella zeae*)
Anthracnose stalk rot	: *Colletotrichum graminicola* (Teleomorph: *Glomerella graminicola*)
Botryodiplodia stalk rot	: *Botryodiplodia theobromae*
Stenocarpella stalk rot	: *Stenocarpella maydis* (syn. *Diplodia maydis*), *S. macrospora*, (syn. *D. macrospora*)

Disease	Causal Organism / Agent
Charcoal stalk rot	: *Macrophomina phaseolina*
Crazy top downy mildew	: *Sclerophthora macrospora*
Brown stripe downy mildew	: *Sclerophthora rayssiae* var. *zeae*
Green ear disease	: *Sclerospora graminicola*
Java downy mildew	: *Peronosclerospora maydis*
Philippine downy mildew	: *Peronosclerospora philippinensis*
Sorghum downy mildew	: *Peronosclerospora sorghi*
Sugarcane downy mildew	: *Peronosclerospora sacchari*
Spontaneum downy mildew	: *P. spontaneae*
Rajasthan downy mildew	: *Peronosclerospora heteropogoni*
Leaf splitting downy mildew	: *P. miscanthi*
Brown spot	: *Physoderma zeae-maydis*
Grey leaf spot	: *Cercospora zeae-maydis*
Common smut	: *Ustilago maydis*
Head smut	: *Sphacelotheca reiliana*
False head smut	: *Ustilaginoidea virens*
Southern corn rust	: *Puccinia polysora*
Common corn rust	: *Puccinia sorghi*
Tropical rust	: *Physopella zeae*
Turcicum leaf blight	: *Exserohilum turcicum* (Teleomorph: *Setosphaeria turcica*)
Maydis leaf blight	: *Bipolaris maydis* (syn. *Helminthosporium maydis*), Teleomorph: *Cochliobolus heterostrophus*
Banded leaf and sheath blight	: *Rhizoctonia solani* (Teleomorph: *Corticium sasakii,* syn. *Thanatephorus cucumeris*)
Ear rots	: *Fusarium graminearum, Fusarium moniliforme* (*syn. F. verticillioides*), *Macrophomina phaseolina, Nigrospora oryzae, Aspergillus* spp. and *Penicillium* spp.
Bacteria	
Bacterial stalk rot	: *Dickeya zeae*
Stewart's wilt	: *Erwinia stewartii* (syn. *Pantoea stewartii*)
Bacterial leaf stripe	: *Pseudomonas rubrilineans,* (syn. *P. avenae, Acidvorax avenae* subsp. *avenae*)
Corn stunt	: *Spiroplasma kunkelii* (syn. corn stunt spiroplasma)
Virus	
	: *Maize chlorotic dwarf virus* (MCDV)
	: *Maize chlorotic mottle virus (MCMV)*
	: *Maize dwarf mosaic virus (MDMV)*
	: *Sugarcane mosaic virus (SCMV)*
	: *Maize streak virus (MSV)*
	: *Maize stripe virus (MStV)*
Disorder/deficiency	
White bud	: Zinc deficiency

26.3 SORGHUM (*SORGHUM BICOLOR* [L.] MOENCH.)

Disease	Causal Organisms/Agent
Fungi and Oomycetes	
Anthracnose (Foliar, head, root and stalk rot)	: *Colletotrichum graminicola* (syn. *Glomerella graminicola*)
Covered/kernel smut	: *Sporisorium sorghi* (syns. *Sphacelotheca sorghi)*
Head Smut	: *Sporisorium reilianum* (syns. *Sphacelotheca reiliana*; *Sporisorium holci-sorghi*)
Long smut	: *Anthracocystis ehrenbergii* (syn. *Sporisorium ehrenbergii*)
Loose kernel smut	: *Sporisorium cruentum* (syn. *Sphacelotheca cruenta*)
Milo disease (Periconia root rot)	: *Periconia circinata*
Pokkah Boeng (Twisted top)	: *Gibberella fujikuroi* (syns. *Fusarium moniliforme* var. *subglutinans*; *G. fujikuroi* var. *subglutinans*), *G. intermedia* (syn. *F. proliferatum*)
Pythium root rot	: *Pythium* spp. (*Pythium arrhenomanes, P. graminicola*)
Banded leaf and sheath blight	: *Rhizoctonia solani* (syn. *Thanatephorus cucumeris*)
Charcoal rot	: *Macrophomina phaseolina*
Crazy top downy mildew	: *Sclerophthora macrospora* (syn. *Sclerospora macrospora*)
Sorghum downy mildew	: *Peronosclerospora sorghi* (syn. *Sclerospora sorghi*)
Zonate leaf spot	: *Gloeocercospora sorghi*
Damping-off and seed rot	: *Aspergillus* spp., *Exserohilum* sp., *Fusarium* spp., *Penicillium* spp., *Pythium* spp., *Rhizoctonia* spp.
Ergot	: *Claviceps africana*, *C. sorghicola*, *Sphacelia sorghi* (syn. *C. sorghi*)
Fusarium head blight, root and stalk rot	: *Fusarium* spp., *Fusarium moniliforme* (syn. *Gibberella fujikuroi*), *F. thapsinum* (syn. *G. thapsina*)
Grain storage mold	: *Aspergillus* spp., *Fusarium andiyazi*, *F. nygamai Penicillium* spp., Other species
Bacteria	
Bacterial leaf blight	: *Acidovorax avenae* subsp. *avenae*
Bacterial leaf spot	: *Pseudomonas syringae* pv. *syringae*
Bacterial leaf streak	: *Xanthomonas vasicola* pv. *holcicola* (syn. *X. campestris pv. holcicola*)
Bacterial leaf stripe	: *Paraburkholderia andropogonis* (syns. *Burkholderia andropogonis; Pseudomonas andropogonis*
Bacterial top and stalk rot	: *Dickeya dadantii* (syn. *Erwinia chrysanthemi*)
Parasitic plant	
Witchweed	: *Striga asiatica, S. forbesi, S. gesnerioides, S. hermonthica*

26.4 BAJRA OR PEARL MILLET (*PENNISETUM GLAUCUM* [L.] R. BR.)

Disease	Causal Organism
Fungi and Oomycets	
Blast	: *Pyricularia grisea*
Downy mildew	: *Sclerospora graminicola*
Ergot	: *Claviceps microcephala*, *C. fusiformis*
Smut	: *Moesiziomyces penicillariae*
Rust	: *Puccinia substriata* var. *indica*
Bacteria	
Leaf spot	: *Xanthomonas penniseti*
Leaf blotch	: *Xanthomonas annamalaiensis*
Nematode	
Reniform nematode	: *Rotylenchulus reniformis*

26.5 GROUNDNUT (*ARACHIS HYPHOGAEA* L.)

Disease	Causal Organism
Fungi	
Collar rot	: *Aspergillus niger*
Stem rot	: *Sclerotium rolfsii*
Aflaroot/yellow mold	: *Aspergillus flavus*
Dry root rot	: *Macrophomina phaseolina*
Early leaf spot	: *Cercospora arachidicola* (Sexual stage: *Mycosphaerella arachidis*)
Late leaf spot	: *Cercosporidium personatum* (Sexual stage: *Mycosphaerella berkeleyi*)
Powdery mildew	: *Oidium arachidis*
Rust	: *Puccinia arachidis*
Fusarium wilt	: *Fusarium oxysporum*
Verticillium wilt	: *Verticillium albo-atrum*, *Verticillium dahlia*
Bacteria	
Bacterial wilt	: *Ralstonia solanacearum*
Nematode	
Kalahasti malady	: *Tylenchorhynchus brevelineatus*
Testa nematode	: *Aphelenchoides arachidis*

26.6 SOYBEAN (*GLYCINE MAX* [L.] MERRIL.)

Disease	Causal Organism/Agents
Fungi and Oomycets	
Rhizoctonia blight/ Rhizoctonia aerial blight and web blight (Rhizoctonia foliar blight)	: *Rhizoctonia solani* (Teleomorph: *Thanatephorus cucumeris*)
Seed and Seedling Rot	: Species of *Fusarium*, *Rhizoctonia*, *Phytophthora*, and *Pythium*
Powdery mildew	: *Erysiphe diffusa*, *E. glycines*
Frogeye leaf spot (Cercospora leaf spot)	: *Cercospora sojina*
Anthracnose	: *Colletotrichum chlorophyti*, *C. coccodes*, *C. gloeosporioides* (syn. *Glomerella cingulata*), *C. graminicola* (syn. *G. graminicola*), *C. incanum*, *C. truncatum*, *Glomerella glycines*
Rust	: *Phakopsora meibomiae* (syn. *Malupa vignae*), *P. pachyrhizi*
Charcoal rot	: *Macrophomina phaseolina*
Bacteria	
Bacterial blight	: *Pseudomonas savastanoi* pv. *glycinea* (syn. *P. syringae* pv. *glycinea*)
Bacterial pustule	: *Xanthomonas axonopodis* pv. *glycines*
Bacterial tan spot	: *Curtobacterium flaccumfaciens* pv. *flaccumfaciens*
Bacterial wilt	: *Curtobacterium flaccumfaciens* pv. *flaccumfaciens*, *Ralstonia solanacearum*
Wildfire	: *Pseudomonas syringae* pv. *tabaci*
Virus	
Soybean mosaic	: *Soybean mosaic virus*
Yellow mosaic	: *Mungbean yellow mosaic virus*
Nematode	
Soybean cyst nematode	: *Heterodera glycines*

26.7 PIGEON PEA/REDGRAM/ARHAR/TUR (*CAJANUS CAJAN* [L.] MILLSP.)

Disease	Causal Organism/Agent
Fungi and Oomycets	
Fusarium wilt	: *Fusarium udum* f. sp. *cajani* (teleomorph: *Gibberella indica*)
Phytophthora blight	: *Phytophthora drechsleri* f. sp. *cajani*
Alternaria blight/leaf spot	: *Alternaria alternate, Alternaria tenuissima*
Powdery mildew	: *Leveillula taurica* (anamorph: Oidiopsis taurica), *Ovulariopsis ellipsospora*

Disease	Causal Organism/Agent
Cercospora leaf spots	: *Mycovellosiella cajani* (*Cercospora cajani*), *Cercospora indica*
Foliar rust	: *Uredo cajani*
Collar rot/soft rot	: *Sclerotium rolfsii*
Bacteria	
Bacterial Leaf Spot and Stem canker	: *Xanthomonas campestris* pv. *cajani*
Virus	
Sterility mosaic	: *Pigeon pea sterility mosaic virus* (PSMV)

26.8 FINGER MILLET (*ELEUSINE CORACANA* [L.] GAERTN.)

Disease	Causal Organism
Fungi and Oomycetes	
Blast	: *Pyricularia oryzae* (teleomorph: *Magnaporthe oryzae*)
Seedling blights and rots	: *Cochiliobolus nodulosum*, *Helminthosporium tetramera*
Downy mildew/green ear/crazy top	: *Sclerophthora macrospora*
Sclerotial root rot (wilt)	: *Sclerotium rolfsii*
Smut	: *Melanopsichium eleusinis*
Bacteria	
Bacterial leaf spot	: *Xanthomonas eleusineae*
Bacterial blight	: *Xanthomonas coracana*

26.9 BLACK GRAM (*VIGNA MUNGO* L.) AND GREEN GRAM/ MUNGBEAN (*VIGNA RADIATA* [L.] WILCZEK)

Disease	Causal Organism/Agent
Fungi and Oomycetes	
Dry root rot, charcoal rot, carbon rot	: *Macrophomina phaseolina*
Wet root rot, web blight	: *Rhizoctonia solani*
Fusarium wilt	: *Fusarium oxysporum*, *F. solani*
Collar rot, southern blight	: *Sclerotium rolfsii*
Phytophthora stem rot	: *Phytophthora vignae* f. sp. *adzukicola* and *P. vignae* f. sp. *vignae*.
Powdery mildew	: *Erysiphe polygoni*, *E. vignae*, *Sphaerotheca phaseoli*, *Podosphaera fusca* pv. *xanthii*
Cercospora leaf spot	: *Cercospora cruenta*, *C. canescens*, *C. caracallae*, *C. kikuchii*
Anthracnose	: *Colletotrichum destructivum* species complex, *C. lindemuthianum*, *C. truncatum*, *C. gloeosporioides*

Disease	Causal Organism/Agent
Alternaria leaf spot	: *Alternaria alternata*
Ascochyta blight	: *Didymella rabiei*
Bacteria	
Halo blight	: *Pseudomonas savastanoi* pv. *phaseolicola*
Bacterial leaf spot	: *Xanthomonas axonopodis* pv. *vignaeradiatae*
Tan spot/bacterial wilt	: *Curtobacterium faccumfaciens* pv. *faccumfaciens*
Virus	
Yellow mosaic	: *Mungbean yellow mosaic virus, Horsegram yellow mosaic virus, Dolichos yellow mosaic virus*
Leaf crinkling	: *Leaf crinkle virus*

26.10 CASTOR (*RICINUS COMMUNIS* L.)

Disease	Causal Organism
Fungi and Oomycetes	
Seedling blight	: *Phytophthora parasitica*
Alternaria blight	: *Alternaria ricini*
Rust	: *Melampsora ricini*
Powdery mildew	: *Leveillula taurica*
Pod rot	: *Cladosporium oxysporium*
Twig blight	: *Glomerealla cingullata*
Leaf spot	: *Cercospora ricinella*
Stem spotting rot	: *Botryodiplodia ricinicola*
Grey mold	: *Botryotinia ricini*
Bacteria	
Bacterial leaf spot	: *Xanthomonas ricinicola*

26.11 TOBACCO (*NICOTIANA TABACUM* L.)

Disease	Causal Organism/Agent
Fungi and Oomycetes	
Black root rot	: *Thielaviopsis basicola*
Damping off	: *Pythium aphanidermatum*
Black shank and leaf blight	: *Phytophthora parasitica* var. *nicotianae, P. nicotianae* var. *nicotianae*
Collar rot	: *Sclerotium rolfsii*
Anthracnose	: *Colletotrichum tabacum, C. destructivum*
Frog-eye spot	: *Cercospora nicotianae*
Brown spot	: *Alternaria alternata*
Powdery mildew	: *Erysiphe cichoracearum*

Disease	Causal Organism/Agent
Ragged leaf spot	: *Phoma exigua* var. *exigua=Ascochyta phaseolorum*
Sore shin and damping-off	: *Rhizoctonia solani* (teleomorph: *Thanatephorus cucumeris*)
Blue mold (downy mildew)	: *Peronospora tabacina*
Bacteria	
Angular leaf spot	: *Pseudomonas syringae* pv. *tabaci*
Wildfire	: *Pseudomonas syringae* pv. *tabaci*
Hollow stalk	: *Erwinia carotovora* subsp. *carotovora*, *E. carotovora* subsp. *atroseptica*
Granville wilt/bacterial wilt	: *Pseudomonas solanacearum*
Leaf gall	: *Rhodococcus fascians*
Virus	
Tobacco mosaic	: *Tobacco mosaic virus*
Tobacco leaf curl	: *Tobacco leaf curl virus*
Tomato spotted wilt	: *Tomato spotted wilt virus*
Tobacco streak	: *Tobacco streak virus*
Tobacco ring spot	: *Tobacco ring spot virus*
Parasitic Higher Plants	
Broomrape	: *Orobanche ramose*, *O. ludoviciana*
Dodder	: *Cuscuta* spp.
Witchweed	: *Striga gesnerioides*
Nematode	
Root-knot	: *Meloidogyne arenaria*, *M. hapla*, *M. incognita*, *M. javanica*

26.12 GUAVA (*PSIDIUM GUAJAVA* L.)

Disease	Causal Organism
Fungi	
Wilt	: *Fusarium oxysporum* f. sp. *psidii, F. solani,* other fungi *Macrophomina phaeseoli*, *Rhizoctonia bataticola*, *Cephalosporium* sp., *Gliocladium roseum*
Anthracnose	: *Colletotrichum psidii*, *Colletotrichum gloeosporioides/ Gloeosporium psidii* (=*Glomerella psidii/Colletotrichum psidii*), *C. gloeosporioides* (teleomorph: *G. cingulata*)
Stem canker	: *Physalospora psidi*
Cercospora leaf spot	: *Cercopsora sawadae*
Grey blight and fruit spot	: *Pestalotia psidii*
Rust	: *Puccinia psidii*
Sooty mold	: *Meliola psidii*
Algae	
Red rust	: *Cephaleuros virescens*

26.13 BANANA (*MUSA PARADISIACA* L.)

Disease	Causal Organism/Agent
Fungi	
Anthracnose	: *Colletotrichum musae*
Fusarium wilt/panama disease	: *Fusarium oxysporum* f. sp. *cubense*
Black leaf streak/Black sigatoka	: *Mycosphaerella fijiensis* (*Pseudocercospora fijiensis*)
Yellow sigatoka	: *Mycosphaerella musicola* (*Pseudocercospora musae*)
Cigar-end rot	: *Verticillium theobromae* and *Trachysphaera fructigena*
Bacteria	
Moko/Bugtok/bacterial wilt	: *Ralstonia solanacearum*
Banana blood disease	: *Ralstonia syzygii* subsp. *celebesensis*
Xanthomonas wilt	: *Xanthomonas campestris* pv. *musacearum*
Bacterial head rot or tip	: *Erwinia carotovora* ssp. *carotovora*; *E. chrysanthemi*
Bacterial rhizome and pseudostem wet rot	: *Dickeya paradisiaca* (formerly *E. chrysanthemi* pv. *paradisiaca*)
Virus	
Banana bunchy top	: *Banana bunchy top virus*
Banana streak disease	: *Banana streak viruses*
Banana bract mosaic disease	: *Banana bract mosaic virus*
Mosaic	: *Cucumber mosaic virus*
Nematode	
Burrowing nematode	: *Radopholus similis*
Root-lesion nematode	: *Pratylenchus* spp.

26.14 PAPAYA (*CARICA PAPAYA* L.)

Disease	Causal Organism/Agent
Fungi and Oomycetes	
Angular leaf spot	: *Leveillula taurica*
Anthracnose	: *Colletotrichum gloeosporioides*
Foot rot	: *Pythium aphanidermatum*, *Pythium ultimum*, *Phytophthora palmivora*, *Phytophthora nicotianae*
Phytophthora blight	: *Phytophthora palmivora*, *Phytophthora nicotianae* var. *parasitica* (=*Phytophthora parasitica*)
Powdery mildew	: *Erysiphe cichoracearum*, *Erysiphe* sp., *Oidium indica*, *Ovulariopsis papaya*, *Sphaerotheca fuliginea*, *S. humuli*
Viral diseases	
Leaf curl	: *Papaya leaf curl virus*
Spotted wilt	: *Tomato spotted wilt tospovirus*
Mosaic	: *Papaya mosaic potexvirus*-1

Disease	Causal Organism/Agent
Papaya ring spot	: *Papaya ringspot potyvirus* (*Papaya ring spot virus*)
Terminal necrosis and wilt	: *Tobacco ringspot nepovirus*
Nematode	
Reniform nematode	: *Rotylenchulus reniformis, R. parvus*
Root-knot nematode	: *Meloidogyne incognita, M. javanica, M. arenaria, M. hapla*

26.15 POMEGRANATE (*PUNICA GRANATUM* L.)

Disease	Causal Organism
Fungi and Oomycetes	
Leaf and fruit spot	: *Pseudocercospora punicae*
Leaf spot and fruit rot	: *Alternaria alternata*
Anthracnose	: *Colletotrichum gloeosporioides*
Wilt	: *Fusarium oxysporum*
Bacteria	
Bacterial leaf and fruit spot	: *Xanthomonas axonopodies* pv. *punicae*

26.16 CRUCIFEROUS VEGETABLES (*BRASSICA* SPP.)

Disease	Causal Organism
Fungi and Oomycetes	
Alternaria leaf spot/Head rot/ Dark leaf spot	: *Alternaria brassicae, A. brassicicola, A. raphani*
Black leg	: *Phoma lingum* (*Leptosphaeria macutans*)
Club root	: *Plasmodiophora brassicac*
Downy mildew	: *Peronospora parasitica*
Fusarium yellows or wilt	: *Fusarium oxysporum conglutinans*
White rust	: *Albugo candida*
Damping-off and wire-stem	: *Pythium* sp. and *Rhizoctonia solani*
White mold/White rot/ Sclerotinia stem rot	: *Sclerotinia sclerotiorum*
Powdery mildew	: *Erysiphe polygoni*
Cercospora leaf spot	: *Cercospora carotae*
Bacteria	
Black rot	: *Xanthomonas campestris* pv. *campestris*
Bacterial soft rot	: *Erwinia carotovora* sp. *carotovora*

26.17 BRINJAL (*SOLANUM MELONGENA* L.)

Disease	Causal Organism
Fungi and Oomycetes	
Damping off	: *Pythium aphanidermatum*, *Pythium indicum*, *Phytophthora parasitica*, *Rhizoctonia solani*, *Sclerotium rolfsii*
Cercospora leaf spot	: *Cercospora solani-melongenae*, *C. solani*
Alternaria leaf spot	: *Alternaria melongenae*, *A. solani*
Phomopsis blight and fruit rot	: *Phomopsis vexans* (*Diaporthe vexans*)
Collar rot	: *Sclerotium rolfsii*
Sclerotinia blight	: *Sclerotinia sclerotiorum*
Verticilium wilt	: *Verticilium dahliae*
Powdery mildew	: *Erysiphe cichoracearum*
Bacteria	
Wilt	: *Ralstonia solanacearum*
Phytoplasma	
Little Leaf	: Phytoplasma
Namatode	
Root knot nematode	: *Meloidogyne incognita*, *M. javanica*
Parasitic plants	
Broom rape	: *Orobanche cernua*, *O. indica*

26.18 TOMATO (*LYCOPERSICON ESCULENTUM* MILL.)

Disease	Causal Organism/Agent
Fungi and Oomycetes	
Damping off	: *Pythium aphanidermatum*
Early blight	: *Alternaria solani*, *A. lternate* f. sp. *lycopersici*
Late blight	: *Phytophthora infestans*
Fusarium wilt	: *Fusarium oxysporum* f. sp. *Lycopersici*
Buck eye rot	: *Phytophthora nicontianae* var. *parasitica*
Powdery mildew	: *Leveillula taurica*
Septoria leaf spot	: *Septoria lycopersici*
Bacteria	
Bacterial wilt	: *Ralstonia solanacearum*
Bacterial stem and fruit canker	: *Clavibacter michiganensis* subsp. *michiganensis*
Bacterial fruits and leaf spots	: *Xanthomonas campestris* pv. *vesicatoria*
Virus	
Tomato mosaic disease	: *Tomato mosaic virus* (ToMV)
Tomato leaf curl disease	: *Tomato leaf curl virus* (ToLCV)
Tomato spotted wilt disease	: *Peanut bud necrosis virus* (PbNV), *Tomato spotted wilt virus* (TSWV)

Disease	Causal Organism/Agent
Nematodes	
Root-knot nematode	: *Meloidogyne* spp.
Reniform nematode	: *Rotylenchulus reniformis*
Physiological disorder	
Blossom-end rot	: Calcium deficiency

26.19 OKRA (*ABELMOSCHUS ESCULENTUS* [L.] MOENCH.)

Disease	Causal Organism/Agent
Fungi and Oomycetes	
Damping off	: *Pythium aphanidermatum*
Fusarium wilt	: *Fusarium oxysporum* f. sp. *vasinfectum*
Phoma canker	: *Phoma exigua*
Powdery mildew	: *Erysiphe cichoracearum*
Leaf spot	: *Cercospora malayensis*
Collor rot	: *Macrophomina phaseolina*
Leaf stem gall	: *Synchytrium hibisci*
Pod and leaf spot	: *Ascochyta abelmoschi*
Virus	
Vein-clearing/Yellow vein mosaic	: *Bhendi yellow vein mosaic virus*
Nematodes	
Root-knot nematode	: *Meloidogyne* spp.
Reniform nematode	: *Rotylenchulus reniformis*

26.20 BEANS (*PHASEOLUS VULGARIS* L.)

Disease	Causal Organism/Agent
Fungi and Oomycetes	
Anthracnose	: *Colletotrichum lindemuthianum* (*Glomerella lindemuthianum*)
Bean root rot	: *Rhizoctonia solani*, *Pythium*, *Fusarium solani*
Powdery mildew	: *Erysiphe polygonii*
Watery soft rot	: *Sclerotinia sclerotiorum*
Rust	: *Uromyces appendiculatus*
Rhizoctonia web blight	: *Rhizoctonia solani* (*Thanatephorus cucumeris*)
Cercospora leaf spot	: *Cercospora cruenta*
Ascochyta leaf spot	: *Ascochyta phaseolorum*

Disease	Causal Organism/Agent
Dry root rot	: *Macrophomina phaseolina*
Fusarium dry root rot	: *Fusarium solani* f. sp. *phaseoli*
Fusarium yellow/wilt	: *Fusarium oxysporum* f. sp. *phaseoli*
Bacteria	
Bacterial blight	: *Xanthomonas campestris* pv. *phaseoli*
Halo blight	: *Pseudomonas syringae* pv. *phaseolicola*
Common blight	: *Xanthomonas campestris* pv. *phaseoli*
Virus	
Bean common mosaic	: *Bean common mosaic virus*
Nematodes	
Root knot	: *Meloidogyne* sp.

26.21 GINGER (*ZINGIBER OFFICINALE* ROSCOE)

Disease	Causal Organism/Agent
Fungi and Oomycetes	
Rhizome rot	: *Armillaria mellea*
Leaf spot	: *Cersospora zingibericola*, *Cercospora zingiberis*
Leaf spot	: *Colletotrichum zingiberis*
Rhizome rot	: *Curvularia lunata*
Dry rhizome rot	: *Diplodia natalensis*
Dry rot of rhizome	: *Macrophomina phaseolina*
Thread blight	: *Pellicularia filamentosa*
Leaf spot	: *Phyllosticta zingiberi*
Rhizome rot	: *Pythium aphanidermatum*, *P. myriotylum*, *P. vexans*
Root rot	: *Pythium zingiberum*
Soft rot of rhizome	: *Pythium myriotylum*
Rhizome rot	: *Rhizopus oryzae*
Basal rot/white rot	: *Sclerotium rolfsii*
Leaf spot	: *Septoria zingiberis*
Fusarium wilt	: *Fusarium oxysporium* f. sp. *zingiberi*
Bacteria	
Bacterial soft rot	: *Erwinia carotovora*
Virus	
Chlorotic fleck	: *Zinger chlorotic fleck virus*

26.22 COLOCASIA OR TARO (*COLOCASIA ESCULENTA* [L.] SCHOTT)

Disease	Causal Organism/Agent
Fungi and Oomycetes	
Phytophthora blight	: *Phytophthora colocasiae*
Leaf spot	: *Alternaria tenuissima*
Storage rot	: *Aspergillus niger, Botryodiploidia theobromae, Fusarium solani, Rhizopus stolonifer*
Leaf spot	: *Cladosporium colocasiae*
Corm rot	: *Cylindrocarpon lichenicola*
Leaf spot	: *Phyllosticya colocasiophilla*
Pythium rot	: *Pythium* sp.
Sclerotium rot	: *Sclerotium rolfsii*
Virus	
	: *Dasheen mosaic virus*
	: *Taro bacilliform virus*
	: *Colocasia bobone disease virus*
	: *Taro vein chlorosis virus*

26.23 COCONUT (*COCOS NUCIFERA* L.)

Disease	Causal Organism/Agent
Fungi and Oomycetes	
Leaf spot/Helminthosporium leaf spot	: *Helminthosporium halodes*
Leaf blight	: *Lasiodiplodia theobromae*
Grey leaf spot	: *Pestalotiopsis palmarum*
Bud rot	: *Phytophthora palmivora*
Stem bleeding	: *Thielaviopsis paradoxa/Ceratocystis paradoxa*
Leaf rot	: *Exserohilum rostratum* and *Colletotrichum gleosporioides*
Basal stem rot (BSR)	: *Ganoderma* spp.
Root wilt	: Phytoplasma
Viroid	
Cadang cadang disease	: *Coconut cadang-cadang viroid* (CCCVd)

26.24 TEA (*CAMELLIA SINENSIS* [L.] O. KUNTZE)

Diseases	Causal Organism
Fungi and Oomycetes	
Birds eye spot	: *Cercospora theae*
Blister blight	: *Exobasidium vexans*
Brown blight	: *Colletotrichum camelliae*
Grey blight	: *Pestalotiopsis theae*
Poria branch canker	: *Poria hypobrunnea*
Black rot	: *Corticium theae, C. invisum*
Charcol stump rot	: *Ustulina zonata*
Brown root rot disease	: *Fomes lamoensis, F. noxius*
Red root disease	: *Poria hypolateritia*
Xylaria rot	: *Xylaria* spp.
Pink disease	: *Pellicularia salmonicola*
Copper blight	: *Guignardia camelliae*
Black rot	: *Corticium invisum*
Root rot	: *Botryodiplodia theobromae*
Stump rot	: *Irpex destruens*
Charcoal rot	: *Ustulina zonata*
Dieback	: *Nectria cinnabarina*
Algae	
Red rust	: *Cephaleuros parasitica, C. mycoides*

26.25 COFFEE (*COFFEA CANEPHORA* L.: ROBUSTA COFFEE; *COFFEA ARABICA* L.: ARABICA COFFEE)

Disease	Causal Organism
Fungi and Oomycetes	
Leaf/orange rust	: *Hemileia vastatrix*
Rust (powdery or grey rust)	: *Hemileia coffeicola*
Black rot or koleroga	: *Corticium salmonicolour*
Brown root	: *Fomes noxius*
Red root	: *Poria hypolateritia*
Sandavery root disease	: *Fusarium oxysporum* f. sp. *coffeae*
Coffee trunk canker	: *Ceratocystis fimbriata*
Berry blotch	: *Cercospora coffeicola*
Brown eye spot disease	: *Mycosphaerella coffeicola*
Coffee berry disease	: *Colletotrichum kahawae*
Coffee wilt disease	: *Fusarium xylarioides*
Coffee bark disease	: *Fusarium stilboides*

Damping off/Collar rot	:	*Pellicularia filamentosa*, *Rhizoctonia solani*
Die back/Anthracnose	:	*Colletotrichum gloeosporioides* (teleomorph = *Glomerella cingulata*), *Colleotrichum kahawae*
Wilt	:	*Ceratocystis fimbriata*, *Fusarium oxysporum*
Armillaria root rot	:	*Armillaria mellea*
Canker	:	*Ceratocystis fimbriata*, *Phomopsis coffeae*
Thread blight	:	*Corticium koleroga*
Dry root rot	:	*Fusarium solani*
Red blister disease (robusta coffee)	:	*Cercospora coffeicola*
Red root rot	:	*Ganoderma philippi*
Pink disease	:	*Corticium salmonicola*
Bacteria		
Bacterial blight	:	*Pseudomonas syringae* pv. *garcae*
Algae		
Algal (red) leaf spot	:	*Cephaleuros virescens*
Nematode		
Root knot	:	*Meloidogyne* spp.

27 Exercises

27.1 MULTIPLE CHOICE QUESTIONS (MCQS)

1. **Which one disease is also known as "rice fever disease"?**
   (a) Blast (b) False smut
   (c) Head scab (d) Bacterial blight
2. **The most damaging blast to the rice crop**
   (a) Leaf blast (b) Collar blast
   (c) Node blast (d) Neck blast
3. **The growth stage of leaf hopper which can transmit *Rice tungro viruses***
   (a) Males (b) Females
   (c) Nymphs (d) All of these
4. **Brown sheath rot of rice is caused by**
   (a) *Pseudomonas fuscovaginae* (b) *Xanthomonas oryzae* pv. *oryzae*
   (c) *Sarocladium oryzae* (d) *Cochiliobolus miyabeans*
5. **The factor which can increase rice susceptibility to the blast disease**
   (a) Drought stress (b) Silicon fertilizers
   (c) Tolerant cultivars (d) Fungicide application
6. ***Pyricularia* causes blast disease in**
   (a) Rice and finger millet (b) Rice and wheat
   (c) Rice and onion (d) Rice, finger millet, and wheat
7. **Disintegration of basal part of stem is called**
   (a) Foot rot (b) Collar rot
   (c) Root rot (d) Stem rot

8. **In rice blast disease, the term blanking means**
   (a) Blackening of leaf lesions
   (b) Non-production of seeds
   (c) Spindle-shaped lesion on seeds
   (d) Abundant sporulation on spots

9. **A combination of rainy weather, strong winds, and temperature (22–26°C) favour the outbreak of**
   (a) Powdery mildew of wheat (b) Bacterial leaf blight of rice
   (c) Downy mildew of maize (d) Root knot disease of okra

10. **Pear-shaped conidia are produced by**
   (a) *Bipolaris oryzae* (b) *Rhizoctonia solani*
   (c) *Pyricularia oryzae* (d) *Ustilaginoidea virens*

11. **Toxin ophiobolin that causes necrotic symptoms in plants produced by**
   (a) *Bipolaris oryzae* (b) *Rhizoctonia solani*
   (c) *Pyricularia oryzae* (d) *Ustilaginoidea virens*

12. **False ergot of rice is caused by**
   (a) *Tilletia barclayana* (b) *Ephelis oryzae*
   (c) *Sarocladium oryzae* (d) *Ustilaginoidea virens*

13. **The conidia of *Pyricularia oryzae* are usually**
   (a) 3-celled (2-septate) (b) 2-celled (1-septate)
   (c) 4-celled (3-septate) (d) 6-celled (5-septate)

14. **The primary source of inoculum for rice blast disease is**
   (a) Infected seeds (b) Infected plant residue
   (c) Collateral hosts (d) All of these

15. **The common practice of cutting the leaf tips of rice seedlings before transplanting can play an important role in the development of**
   (a) Bacterial blight (b) Sheath blight
   (c) Blast (d) Brown spot

16. **Bacterial leaf blight of rice is caused by.........while bacterial leaf streak is caused by...............**
   (a) *Xanthomonas campestris* pv. *malvacearum, Xanthomonas oryzae* pv. *oryzae*
   (b) *Xanthomonas oryzae* pv. *oryzae, Xanthomonas oryzae* pv. *oryzicola*
   (c) *Xanthomonas oryzae* pv. *oryzicola, Xanthomonas oryzae* pv. *oryzae*
   (d) *Xanthomonas axonopodis* pv. *citri, Xanthomonas campestris* pv. *vesicatoria*

17. **Blast disease of rice is caused by**
   (a) *Fusarium fujikuroi* (b) *Pyricularia oryzae*
   (c) *Sclerotium oryzae* (d) *Alternaria padwickii*

18. **The conidia of *Pyricularia oryzae* overwinter on**
   (a) Collateral hosts (b) Infected crop refuse
   (c) Infected seeds (d) All of these

19. Symptom 'Kresek' phase is associated with the
(a) Downy mildew of maize (b) Ergot of sorghum
(c) Blast disease of rice (d) Bacterial blight of rice

20. Resistant variety/cultivar having polygenic resistance including for bacterial leaf blight of rice is
(a) Improved Basmati-1 (b) NH-56
(c) Jyothi (d) All of these

21. "A rattle-snake skin pattern" on rice leaf is the characteristic symptom of
(a) Blast (b) Bacterial leaf blight
(c) Udbatta (d) Sheath blight

22. The disease which replaces the whole rice grain into a blackish spore ball
(a) Blast (b) False smut
(c) Head scab (d) Flag smut

23. False smut of rice is
(a) An epidemic disease (b) An endemic disease
(c) A sporadic disease (d) A pandemic disease

24. False smut pathogen of rice survives through
(a) Sclerotia (b) Infected seed
(c) Chlamydospores (d) All of these

25. Mycotoxin "ustiloxins" which are toxic to plants and animals and inhibit mitosis by interfering with microtubule function, are produced by
(a) *Aspergillus flavus* (b) *Ustilaginoidea virens*
(c) *Fusarium graminearum* (d) *Bipolaris oryzae*

26. False smut of rice is
(a) Soilborne (b) Seedborne
(c) Airborne (d) All of these

27. Popularly false smut of rice is also known as
(a) Haldi rog (b) Lakshmi rog
(c) Kandua rog (d) All of these

28. The causal agent of false smut of rice
(a) *Ustilaginoidea virens* (b) *Rhizoctonia solani*
(c) *Ustilago maydis* (d) *Urocystis agropyri*

29. Bakanae disease of rice is
(a) Seedborne (b) Soilborne
(c) Airborne (d) Both (a) and (b)

30. Tungro disease of rice is caused by
(a) Nematode (b) Phytoplasma
(c) Virus (d) Fastidious bacteria

31. Tungro disease of rice is transmitted by
(a) Thrips (b) Aphid
(c) Leaf hopper (d) Plant hopper

32. Khaira disease of rice cannot be corrected by applying
(a) CAN (b) $ZnSO_4$
(c) ZnO (d) Zinc-coated urea

33. Plant disease related to the discovery of Gibbrellins
(a) False smut of rice (b) Foolish seedling of rice
(c) Crown gall of apple (d) Common smut of maize

34. Which of the following phase of rice blast disease is most damaging?
(a) Leaf blast (b) Collar blast
(c) Panical blast (d) Neck blast

35. Which of the following is a polycyclic disease?
(a) Rice blast (b) Wheat rust
(c) Potato late blight (d) All of these

36. A forecasting model *"Epi-Bla"* has been developed in India for
(a) Rice blast (b) Wheat rust
(c) Late blight of potato (d) Downy mildew of grapes

37. Rice blast disease is favoured by
(a) Long periods of leaf wetness (for 6–8 hours)
(b) High relative humidity (80%–100%)
(c) Optimum temperature (25°C–28°C)
(d) All of these

38. Which of the following is a blast resistant variety of rice?
(a) Krishna Hamsa (b) Nellore Mahsuri
(c) TKM 10 (d) All of these

39. Fungicide used for the management of rice blast disease
(a) Ediphenphos (b) Tricyclazole
(c) Kasugamycin (d) All of these

40. Brown spot of rice caused by
(a) *Pyricularia oryzae* (b) *Bipolaris oryzae*
(c) *Sarocladium oryzae* (d) *Rhizoctonia solani*

41. Fungus that grows on infected rice leaves and sheaths as microscopic runner hyphae
(a) *Pyricularia oryzae* (b) *Bipolaris oryzae*
(c) *Ustilaginoidea virens* (d) *Rhizoctonia solani*

42. *Rhizoctonia solani* survives through infected
(a) Plant debris (b) Seed
(c) Weed hosts (d) All of these

43. Factors that favour the infection spreads most quickly and development of sheath blight of rice
(a) High temperature (28°C–32°C) (b) High relative humidity (85%–100%)
(c) Dense plant stands (d) All of these

44. *Villosiclava* is a sexual stage of

(a) *Rhizoctonia solani* (b) *Ustilaginoidea virens*
(c) *Macrophomina phaseolina* (d) *Colletotrichum falcatum*

45. "Tungro" means

(a) Yellowing (b) Degenerated growth
(c) Severe curling (d) Severe necrosis

46. Wilting syndrome "kersek" occurs in

(a) Bacterial blight of cotton (b) Bacterial blight of rice
(c) Loose smut of wheat (d) Yellow ear rot of wheat

47. Fungus that transforms individual rice grains into large greenish velvety spore-balls is

(a) *Ustilago maydis* (b) *Entyloma oryzae*
(c) *Ustilago tritici* (d) *Ustilaginoidea virens*

48. Leaf smut of rice is caused by

(a) *Ustilago maydis* (b) *Entyloma oryzae*
(c) *Ustilago tritici* (d) *Ustilaginoidea virens*

49. Rice blast is also known as

(a) Rotten neck (b) Rice fever
(c) Brusone (d) All of these

50. The carrier of ergot of pearl millet is

(a) Seed (b) Air
(c) Insect (d) Both (a) and (c)

51. The term "stalk rot" is widely used to describe the

(a) Premature senescence and deterioration of basal stalk
(b) Premature plant death
(c) Soft, often broken, lower stalk internodes
(d) All of these

52. In rice leaves, *Xanthomonas oryzae* pv. *oryzae* enter though...........while *Xanthomonas oryzae* pv. *oryzicola* enter through the.............

(a) Stomata, hydathodes (b) Lenticels, stomata
(c) Hydathodes, stomata (d) Leaf hairs, hydathodes

53. Which one of the following is NOT a pathogen of maize crops?

(a) *Physoderma maydis* (b) *Colletotrichum graminicola*
(c) *Macrophomina phaseolina* (d) *Puccinia graminis tritici*

54. Which of the following fungus produces asexual fruiting structures that contain tiny, dark, whisker-like appendages, called setae, on infected tissue of maize?

(a) *Physoderma maydis* (b) *Colletotrichum graminicola*
(c) *Macrophomina phaseolina* (d) *Rhizoctonia solani*

55. ***Pythium aphanidermatum*** **survives in infested soil residue as**
(a) Mycelia (b) Sporangia
(c) Oospores (d) All of these

56. Pythium stalk rot of maize is more prevalent and infection is favoured by
(a) Wet and high humid weather
(b) Extended periods of high temperatures (25°C–35°C)
(c) High crop density and high levels of nitrogen fertilizer
(d) All of these

57. Inside the maize stalks numerous tiny microsclerotia are present that give a speckled appearance, similar to a silvery black charcoal dust produced by
(a) *Colletotrichum graminicola* (b) *Pythium aphanidermatum*
(c) *Macrophomina phaseolina* (d) *Erwinia chrysanthemi* pv. *zeae*

58. Which one of the following pairs is incorrect?
(a) Charcoal rot of maize—Soilborne disease
(b) Fusarium stalk rot of maize—Seedborne disease
(c) False smut of rice—Seedborne disease
(d) Bacterial blight of rice—Airborne

59. The first record of downy mildew on maize in India was made by
(a) E.J. Butler (1913) (b) Y.L. Nene (1968)
(c) R.S. Singh (1972) (d) G. Rangaswami (1975)

60. Typical "*half-diseased leaf*" symptoms appear in
(a) Sugarcane downy mildew of maize
(b) Brown stripe downy mildew of maize
(c) Crazy top of maize
(d) Rajasthan downy mildew of maize

61. Downy mildew pathogen of maize that is seed and soilborne
(a) Crazy top of maize
(b) Brown stripe downy mildew of maize
(c) Sugarcane downy mildew
(d) Philippine downy mildew of maize

62. Downy mildews of maize can be effectively managed by
(a) Following long crop rotation with pulses
(b) Rouging out infected plants
(c) Treating seed with Metalaxyl
(d) All of these

63. Brown spot of maize is caused by
(a) *Cercospora zeae-maydis* (b) *Sclerophthora macrospora*
(c) *Colletotrichum graminicola* (d) *Physoderma maydis*

64. Which one pair of sorghum diseases is incorrect?
(a) Covered smut—*Sporisorium sorghi*
(b) Loose smut—*Ustilago nuda*

(c) Long smut—*Sporisorium ehrenbergii*
(d) Head smut—*Sporisorium reilianum*

65. The acervuli with setae are diagnostic features of
(a) Anthracnose of sorghum
(b) Anthracnose of chilli
(c) Anthracnose of grapes
(d) All of these

66. *Physoderma maydis* (brown spot of maize) is
(a) An obligate parasite (b) A saprophyte
(c) A facultative parasite (d) A necrotroph

67. Infection hyphae of *Physoderma maydis*, causing a brown spot of maize, expand within the host cells to form several enlarged storage cells called
(a) Vesicles (b) Sammelzellen
(c) Haustoria (d) Arbuscules

68. Brown spot of maize caused by *Physoderma maydis* is favoured by
(a) High temperature (28°C–30°C) and abundant moisture
(b) High temperature (28°C–30°C) and low moisture
(c) Low temperature (15°C–20°C) and abundant moisture
(d) Low temperature (15°C–20°C) and low moisture

69. Best option for the management of brown spot of maize
(a) Field sanitation (b) Crop rotation
(c) Spray any copper fungicide (d) All of these

70. Grey leaf spot of maize is caused by
(a) *Bipolaris zeicola* (b) *Cercospora zeae-maydis*
(c) *Physoderma maydis* (d) *Alternaria solani*

71. Most of the sorghum mold-causing fungi are disseminated by
(a) Seedborne spores (b) Soil-borne spores
(c) Airborne spores (d) All of these

72. "Green ear" disease of pearl millet is caused by
(a) *Sclerospora graminicola* (b) *Peronosclerospora heteropogoni*
(c) *Sclerospora sacchari* (d) *Peronosclerospora sorghi*

73. "Half-leaf symptom" is associated with the disease
(a) Powdery mildew (b) Downey mildew
(c) Leaf spot (d) Wilt

74. Which of the following pathogen is an obligate parasite?
(a) Downy mildew (b) Powdery mildew
(c) Rust (d) All of these

75. The phanerogamic parasite with the sorghum crop
(a) *Striga* (b) *Orobanche*
(c) *Loranthus* (d) *Cuscuta*

76. Transformation of floral parts into leafy structures is called

(a) Green ear (b) Phyllody
(c) Leaf curl (d) Ergot

77. Green ear disease of pearl millet can be managed by

(a) Use of disease-free seed
(b) Foliar spray of Ridomil MZ 72 @ 3 g/L of water
(c) Crop rotation with nonhost crops
(d) All of these

78. The loss of turgidity and drooping of leaves or shoots is called

(a) Wilt (b) Die-back
(c) Decline (d) Root rot

79. Green ear is the characteristic symptom produced by

(a) *Pseudoperonospora cubens* (b) *Plasmopora viticola*
(c) *Sclerospora graminicola* (d) *Claviceps purpurea*

80. Upon germination sclerotial bodies of *Claviceps microcephala* produce

(a) Conidia in pycnidia (b) Ascospores in apothecia
(c) Ascospores in perithecia (d) Conidia in acervuli

81. Seed treatment with a fungicide is NOT useful against

(a) Head smut of sorghum (b) Smut of ragi
(c) Loose smut of wheat (d) Grain smut of sorghum

82. Tikka disease of groundnut is also known as

(a) Peanut cercosporosis
(b) Mycosphaerella leaf spot
(c) Viruela
(d) All of these

83. Early leaf spot groundnut is caused by...........and late leaf spot is caused by.........

(a) *Cercospera arachidicola*, *Cercosporidium personatum*
(b) *Cercosporidium personatum*, *Cercospera arachidicola*
(c) *Cercospora fragariae*, *Cercospora melongena*
(d) *Cercospora canescens*, *Cercospora beticola*

84. Resistant/tolerant variety to tikka disease of groundnut

(a) Vemana (b) Somnath
(c) Apoorva (d) All of these

85. The ergot of pearl millet was first reported from

(a) North India (b) South India
(c) West Bengal (d) Assam

86. The first symptom of ergot of pearl millet appears as

(a) Black moldy growth on panicles
(b) Droplets of "honeydew" on infected panicles
(c) White powdery growth on panicles
(d) Dieback of panicles

87. Leaf shredding and green ear symptoms are noticed in
(a) Downy mildew (b) Wilt
(c) Rusts (d) Smuts

88. Grain smut of sorghum is caused by
(a) *Sphacelotheca sorghi* (b) *Sphacelotheca cruenta*
(c) *Sphacelotheca reiliana* (d) *Tolyposporium ehrenbergii*

89. Sclerotia of *Claviceps fusiformis* germinate to form
(a) Perithecia (b) Apothecia
(c) Cleistothecia (d) Acervuli

90. The primary disease cycle of ergot of bajra begins with the
(a) Conidia (b) Ascospores
(c) Mycelia (d) Sclerotia

91. Conidia of *Claviceps fusiformis* are spread from plant to plant by
(a) Houseflies (b) Wind
(c) Rainsplash (d) All of these

92. In ergot disease of pearl millet, the honeydew droplets contain mycelial mass, conidia, and sugary liquid dry out and transform into sclerotia within
(a) 1–5 days (b) 5–10 days
(c) 10–15 days (d) 20–25 days

93. Infection and spread of ergot disease of pearl millet is favoured by
(a) Drizzling rain and wind
(b) High humidity (more than 80%)
(c) Moderate temperature (20°C–30°C)
(d) All of these

94. Infection of ergot disease of pearl millet occurs during
(a) Panicle initiation stage (b) Seed germination
(c) Grain filling stage (d) Flowering stage

95. Green ear or downy mildew of pearl millet was first time reported in India by
(a) K.C. Mehta (b) E.J. Butler
(c) B.B. Mundkar (d) R. Prasad

96. For the effective management of wilt, pigeon pea should be intercropped with
(a) Maize (b) Pearl millet
(c) Sorghum (d) Mungbean

97. Grain smut of sorghum is caused by
(a) *Sphacelotheca reiliana* (b) *Sphacelotheca cruenta*
(c) *Sphacelotheca sorghi* (d) *Tolyposporium erenbergii*

98. Downy mildew disease of pearl millet is primarily a
(a) Seedborne (b) Airborne
(c) Soilborne (d) Waterborne

99. The management of ergot of bajra may be possible by adopting of
(a) Long crop rotation
(b) Keeping seed in 10% salt solution
(c) Spray of Azoxystrobin and Benlate
(d) All of these

100. Infection of smut of maize occurs
(a) After flowering (b) Before flowering
(c) During reproductive stage (d) During vegetative stage

101. Leaf splitting downy mildew of maize is caused by
(a) *Sclerospora sorghi* (b) *Sclerophthora macrospora*
(c) *Sclerospora sacchari* (d) *Peronospora miscanthi*

102. Downy mildew symptoms of maize are masked by
(a) Infestation of shoot fly
(b) Infection of *Maize streak virus*
(c) Infection of *Maize dwarf mosaic virus*
(d) Deficiency of Zinc

103. Downy mildew of bajra is caused by
(a) *Sclerospora sorghi* (b) *Sclerospora graminicola*
(c) *Sclerospora sacchari* (d) *Peronospora parasitica*

104. Leaf wetness required for infection by zoospores of brown stripe downy mildew pathogen of maize
(a) 6 hours (b) 24 hours
(c) 12 hours (d) 72 hours

105. In India, aerial blight or web blight of soybean was first reported in
(a) 1967 (b) 1975
(c) 1981 (d) 1993

106. Aerial blight or web blight of soybean is caused by
(a) *Rhizoctonia bataticola* (b) *Rhizoctonia solani*
(c) *Alternaria solani* (d) *Macrophomina phaseolina*

107. The perfect stage of *Rhizoctonia solani* produces
(a) Conidia (b) Zoospores
(c) Basidiospores (d) None of these

108. During warm, wet weather, mycelium spreads extensively on the surface of soybean plants, forming localised mats of "webbed" foliage which is caused by
(a) *Macrophomina phaseolina* (b) *Rhizoctonia solani*
(c) Spiders (d) Mites

109. Aerial blight or web blight of soybean is managed by
(a) Destroying infected stubble (b) Avoiding dense planting
(c) Spraying Carbendazim (d) All of these

110. Bacterial pustule of soybean is caused by
(a) *Pseudomonas savastanoi* pv. *glycinea*

(b) *Xanthomonas axonopodis* pv. *glycines*
(c) *Pseudomonas syringae* pv. *maculicola*
(d) *Xanthomonas oryzae* pv. *oryzae*

111. Which of the following statement is correct about Cercospora leaf spots of groundnut?
(a) Late leaf spots are mostly produced on the lower surface
(b) In early leaf spots, fungal sporulation mostly occurs on the upper surface
(c) Concentric rings on spots are obvious in late leaf spots
(d) All of these

112. "Webbed foliage" is a typical symptom of
(a) Charcol rot of soybean
(b) Rhizoctonia foliar blight of soybean
(c) Downy mildew of maize
(d) Ergot of pearl millet

113. The soybean pustule causing bacterium survives in
(a) Infected crop residue
(b) Infected seed
(c) Rhizosphere of wheat roots
(d) All of these

114. The soybean pustule causing bacterium disseminated through
(a) Wind-splashed rain
(b) Overhead-irrigation
(c) Cultural operations during wet foliage
(d) All of these

115. The optimum temperature for soybean pustule development is
(a) 0°C–5°C (b) 10°C–13°C
(c) 20°C–25°C (d) 30°C–33°C

116. Chlorosis is
(a) Systemic toxemia (b) Yellowing
(c) Degradation of chlorophyll (d) All of these

117. Bacterial toxin coronatine interferes with
(a) Metabolic activities
(b) Transpiration process
(c) Biosynthesis of growth regulators
(d) Photosynthesis and chlorophyll production

118. Bacterial blight of soybean is favoured by
(a) Continuous soybean cropping
(b) Zero-tillage production systems
(c) Fields planted with infected seed lots
(d) All of these

119. Bacterial blight of soybean is managed by
(a) Use of pathogen-free seed for sowing
(b) Crop rotation with maize and sorghum

(c) Spray of Streptocycline + Copper oxychloride
(d) All of these

120. Bioagent used for the inhibition of local and systemic development of bacterial blight of soybean
(a) *Bdellovibrio bacteriovorus* (b) *Bacillus subtilis*
(c) *Beauveria bassiana* (d) *Trichoderma harzianum*

121. Which of the following characters distinguishes wildfire disease from other bacterial diseases of soybean?
(a) Dead necrotic areas (b) Yellow halo
(c) Raised margins (d) Bacterial ooze

122. Wildfire disease of soybean is caused by
(a) *Pseudomonas fluorescens*
(b) *Pseudomonas syringae* subsp. *tabaci*
(c) *Pseudomonas savastanoi* pv. *glycinea*
(d) *Xanthomonas axonopodis* pv. *glycines*

123. Which of the following bacterial disease of soybean is seedborne?
(a) Bacterial pustule (b) Bacterial Blight
(c) Wildfire (d) All of these

124. Which of the following pathogen can attack seed and seedlings of soybean, causing pre- or post-emergence damping off or both?
(a) *Cercospora sojina* (b) *Cercospora kikuchii*
(c) *Macrophomina phaseolina* (d) All of these

125. Most common pathogen tends to be associated with seed and seedlings rot of soybean
(a) *Fusarium* and *Rhizoctonia* (b) *Phytophthora*, and *Pythium*
(c) Both (a) and (b) (d) None of these

126. Twisted maize stalk during lodging can be seen in
(a) Gibberella stalk rot (b) Pythium stalk rot
(c) Diplodia stalk rot (d) Charcoal rot

127. Which of the following statement is correct about *Pseudomonas syringae* pv. *tabaci*?
(a) It is hemi-biotrophic bacterial pathogen
(b) It causes wildfire disease in many plants
(c) It produces tabtoxin
(d) All of these

128. Genome of the *Soybean mosaic virus* is
(a) ssRNA molecule (b) dsRNA molecule
(c) ssDNA molecule (d) dsDNA molecule

129. *Soybean mosaic virus* is transmitted by
(a) Aphids (b) Cell sap
(c) Infected seed (d) All of these

130. Early infection of *Soybean mosaic virus* in soybean can
(a) Reduces pod set (b) Increases seed coat mottling
(c) Reduces seed size and weight (d) All of these

131. *Soybean mosaic virus* transmitting aphids become viruliferous by feeding on
(a) Stem tips (b) Infected stems
(c) Old or young leaves (d) All of these

132. Which of the following is a typical symptom of a yellow mosaic of soybean?
(a) Yellow mosaic (b) Leaf mottling and crinkling
(c) Stunted growth of plants (d) All of these

133. *Mungbean yellow mosaic virus* belongs to
(a) Caulimoviridae (b) Tymoviridae
(c) Potyviridae (d) Geminiviridae

134. *Mungbean yellow mosaic virus* is transmitted by
(a) Whitefly (b) Aphid
(c) Nematode (d) Thrips

135. *Mungbean yellow mosaic virus* is NOT transmitted by
(a) Whitefly (b) Grafting
(c) Sap and seed (d) All of these

136. Soybean mosaic disease is managed by
(a) Use of SMV-free seeds
(b) Use of resistant varieties
(c) Spray of pyrethroid or organophosphate
(d) All of these

137. Purple seed stain of soybean is caused by
(a) *Diaporthe phaseolorum* (b) *Rhizoctonia solani*
(c) *Soybean mosaic virus* (d) *Cercospora kikuchii*

138. Pigeon pea disease which is known as "green plague"
(a) Wilt (b) Sterility mosaic
(c) Phytophthora blight (d) Charcoal rot

139. The sterility mosaic of pigeon pea was first described in India in
(a) 1921 (b) 1956
(c) 1972 (d) 1931

140. Sterility mosaic disease of pigeon pea is characterised by
(a) Bushy and pale green appearance of the plants
(b) Complete or partial cessation of flower production
(c) Mosaic or chlorotic ringspot symptoms on the leaves
(d) All of these

141. Symptoms of sterility mosaic disease of pigeon pea are often masked in
(a) Older plants (b) Young plants
(c) Adult plants (d) All of these

142. Virion of *Pigeon pea sterility mosaic virus* is
(a) Icosahedral geminate (b) Complex
(c) Thread-like flexuous (d) None of these

143. Genome of *Pigeon pea sterility mosaic virus* is
(a) ssRNA molecule (b) dsRNA molecule
(c) ssDNA molecule (d) dsDNA molecule

144. Primary source of inoculum of *Pigeon pea sterility mosaic virus* is
(a) Ratooned pigeon pea
(b) Diseased plants left in the field
(c) Perennial pigeon pea and its wild relatives
(d) All of these

145. *Pigeon pea sterility mosaic virus* is NOT transmitted by
(a) Seed (b) Plant sap
(c) Thrips (d) All of these

146. Minimum number of eriophyid mites that are enough to transmit the *Pigeon pea sterility mosaic virus*
(a) 1 (b) 3
(c) 5 (d) 8

147. The growth stage of eriophyid mite which is efficient to transmit the *Pigeon pea sterility mosaic virus*
(a) Nymphs (b) Adults
(c) Eggs (d) Both (a) and (b)

148. *Pigeon pea sterility mosaic virus* is transmitted through
(a) Seed (b) Plant sap
(c) Thrips (d) Eriophyid mite

149. Pigeon pea sterility mosaic disease can be managed by
(a) Rouging out the infected plants
(b) Spray of Fenazaquin or Dimethoate
(c) Crop rotation to reduce the mite population
(d) All of these

150. Stem blight and canker of pigeon pea is caused by
(a) *Rhizoctonia solani*
(b) *Pseudomonas syringae* pv. *phaseolicola*
(c) *Phytophthora palmivora*
(d) *Phytophthora drechsleri* f.sp. *cajani*

151. The incidence of Phytophthora blight of pigeon pea increases in
(a) High doses of nitrogen
(b) Absence of potassium
(c) High humidity and a temperature (28°C–32°C)
(d) All of these

152. Practice that will NOT be useful for the management of Phytophthora blight of pigeon pea
(a) Seed treatment with Ridomil
(b) Crop rotation with leguminous crops

(c) Use of potassium-based fertilizers
(d) Use of pathogen-free seeds for planting

153. In India, the vascular wilt of pigeon pea was first recorded by
(a) Sarkar et al. (2000) (b) Vidyasekaran et al. (1997)
(c) Kotasthane et al. (1987) (d) Butler (1906)

154. Soybean cyst nematode is
(a) *Heterodera schachtii* (b) *Hetrerodera cajani*
(c) *Heterodera glycines* (d) *Globodera rostochiensis*

155. Pigeon pea cyst nematode is
(a) *Heterodera avenae* (b) *Hetrerodera cajani*
(c) *Heterodera glycines* (d) *Globodera rostochiensis*

156. Vascular wilt of pigeon pea is caused by
(a) *Fusarium udum* f.sp. *crotalariae*
(b) *Fusarium oxysporum* f.sp. *lentis*
(c) *Fusarium oxysporum* f.sp. *ciceri*
(d) *Fusarium udum* f.sp. *cajani*

157. The symptoms of vascular wilt of pigeon pea are characterised by
(a) Droopping and loss of turgidity
(b) Slight interveinal chlorosis on leaves
(c) A purple band on the stem
(d) All of these

158. Pigeon pea wilt pathogen survives in
(a) Soil (b) Plant stubbles
(c) Seeds (d) All of these

159. Pigeon pea wilt pathogen produces
(a) Microconidia (b) Macroconidia
(c) Chlamydospores (d) All of these

160. Which of the following play an important role in the spread of pigeon pea wilt disease by injuring the plants and also affecting their growth?
(a) Mites (b) Spotted pod borer
(c) Pod fly (d) Nematodes

161. Which of the following is associated with the breakdown of resistance by retarding the accumulation of cajanol an enzyme responsible for resistance in pigeon pea cultivar?
(a) Spotted pod borer (b) High temperature
(c) Cyst nematodes (d) Root-knot nematodes

162. Most effective and practical solution for the management of pigeon pea wilt is mixed or intercropping with
(a) Sorghum (b) Wheat
(c) Maize (d) Mustard

163. Recommended crops for rotations to manage the pigeon pea wilt is
(a) Tobacco and sorghum (b) Pearl millet and cotton
(c) Resistant pigeon pea genotypes (d) All of these

164. Fusarium wilt-resistant genotype or variety of pigeon pea is
(a) Narendra Arhar-1 (b) BDN2
(c) MA3 (d) All of these

165. Blast of ragi was recorded first in India by
(a) McRae (1920) (b) Butler (1906)
(c) Basu and SenGupta (2001) (d) Joshi and Singh (1969)

166. Blast of ragi was recorded first in India from
(a) Tamil Nadu (b) Karnataka
(c) Uttarakhand (d) Punjab

167. Initial inoculum for blast disease of ragi comes from
(a) Infected seeds (b) Weeds or collateral hosts
(c) Airborne conidia (d) All of these

168. Blast disease of ragi is favoured by the temperature of
(a) 10°C–15°C (b) 15°C–25°C
(c) 25°C–30°C (d) 30°C–35°C

169. Brown spot of finger millet is caused by
(a) *Cercospora eleusinis* (b) *Helminthosporium oryzae*
(c) *Pyricularia grisea* (d) *Drechslera nodulosa*

170. Cercospora leaf spot of mungbean and urdbean is caused by
(a) *Cercospora cruenta* (b) *Cercospora canescens*
(c) *Cercospora kikuchii* (d) All of these

171. Secondary infection of Cercospora leaf spot of mungbean and urdbean is initiated by conidia under
(a) Warm and wet conditions (b) Cool and wet conditions
(c) Warm and dry conditions (d) Cool and dry conditions

172. "Oil globules" are found in conidia of
(a) *Fusarium oxysporium* (b) *Colletotrichum lindemuthianum*
(c) *Alternaria solani* (d) *Helminthosporium oryzae*

173. Anthracnose of mungbean and urdbean is can be managed by hot water treatment of seed at
(a) 50°C for 10 minutes (b) 52°C for 10 minutes
(c) 58°C for 15 minnutes (d) 65°C for 15 minutes

174. In India, MYMV was first reported by
(a) Ahlawat (2002) (b) Nariani (1960)
(c) Tripathi (1996) (d) Kulkarni (1986)

175. Halo blight of mungbean is caused by
(a) *Pseudomonas savastanoi* pv. *phaseolicola*
(b) *Xanthomonas axonopodis* pv. *vignaeradiatae*
(c) *Curtobacterium flaccumfaciens* pv. *flaccumfaciens*
(d) None of these

176. Select the INCORRECT pair of mungbean diseases and their pathogen
(a) Bacterial leaf spot—*Xanthomonas axonopodis* pv. *vignaeradiatae*
(b) Tan spot/bacterial wilt—*Curtobacterium flaccumfaciens* pv. *flaccumfaciens*
(c) Halo blight—*Pseudomonas savastanoi* pv. *phaseolicola*
(d) Powdery mildew—*Podosphaera leucotricha*

177. *Macrophomina phaseolina* causes
(a) Dry root rot (b) Charcoal rot
(c) Carbon rot (d) All of these

178. The presence of white to brown cysts projecting on the root surface is a characteristic symptom of
(a) *Longidorus* (b) *Globodera*
(c) *Meloidogyne* (d) *Trichodorus*

179. The characteristic symptom of yellow mosaic disease of mungbean and urdbean is
(a) Malformation (b) Leaf yellowing
(c) Stunting of plants (d) All of these

180. *Mungbean yellow mosaic India virus* is NOT transmitted through
(a) Sap (b) Pollen
(c) Seed (d) All of these

181. Phytophthora blight disease of the castor oil plant discovered by
(a) E.J. Butler (1906) (b) J.F. Dustur (1913)
(c) C.H. Persoon (1801) (d) R.S. Singh (1981)

182. Who establish a new species of *Phytophthora* and named *P. parasitica?*
(a) E.J. Butler (b) J.F. Dustur
(c) Y.L. Nene (d) R.S. Singh

183. Which one of the following spores is produced by *Phytophthora parasitica*?
(a) Zoospores (b) Oospores
(c) Chlamydospores (d) All of these

184. Black shank disease of tobacco is caused by
(a) *Phytophthora parasitica* (b) *Phytophthora nicotianae*
(c) *Thielaviopsis basicola* (b) *Rhizoctonia solani*

185. Which one of the following is a polycyclic disease?
(a) Wheat rust (b) Black shank of tobacco
(c) Late blight of potato (d) All of these

186. Black root rot disease of tobacco is also called
(a) Maricume radicale (b) Root rot
(c) Thielavia root rot (d) All of these

187. Black root rot disease of tobacco is caused by
(a) *Thielaviopsis basicola* (b) *Rhizoctonia solani*
(c) *Colletotrichum tabacum* (d) *Phytophthora nicotianae*

188. Sore shin disease of tobacco is caused by
(a) *Thielaviopsis basicola* (b) *Rhizoctonia solani*
(c) *Colletotrichum tabacum* (d) *Phytophthora nicotianae*

189. Fungus that produces endoconidia and aleuriospores
(a) *Thielaviopsis basicola* (b) *Rhizoctonia solani*
(c) *Colletotrichum capsici* (d) *Fusarium oxysporum*

190. Which of the following plant pathogen is a hemibiotrophic pathogen?
(a) *Phytophthora infestans* (b) *Thielaviopsis basicola*
(c) *Magnaporthe oryzae* (d) All of these

191. In a necrotrophic lifestyle plant pathogen secrets
(a) Cell wall degrading enzymes (b) Oxalic acid
(c) Toxins (d) All of these

192. Select the wrong pair
(a) Mosaic—Light and dark green patches
(b) Mottling—Patterns of green and yellow areas
(c) Necrosis—Drooping and loss of turgidity
(d) Puckering—Pouch-like development of green parts of leaves

193. Black root rot disease of tobacco is more severe at pH
(a) Above 5.6 (b) Below 5.2
(c) 4.2 (d) 5.0

194. Black root rot disease of tobacco effectively managed by
(a) Growing resistant cultivars
(b) Avoiding leguminous plants as cover crops
(c) Maintaining soil pH below 5.2
(d) All of these

195. The symptom on tobacco produced by *Tobacco mosaic virus* under hot weather
(a) Mosaic burn (b) Mottling and necrosis
(c) Shoestring effect (d) Dark-green blisters

196. A common source of new infection *Tobacco mosaic virus*
(a) Tobacco seed (b) Air-dried tobacco
(c) Insect vector (d) All of these

197. TMV is transmitted most rapidly and very easily by
(a) Sap (b) Contaminated farm implements
(c) Air-dried tobacco (d) All of these

198. The purified virions of TMV can remain infectious when stored at 4°C in the laboratory for
(a) 5 years (b) 10 years
(c) 25 years (d) 50 years

199. Guava wilt disease was first reported in India by
(a) Das Gupta and Rai (1947) (b) Bapat and Shah (1997)
(c) Basu and SenGupta (2004) (d) Krishnamurthy and Gnanamanickam (1982)

200. In slow wilt of guava, plants take............for complete wilting.

(a) 3 days (b) 5 days
(c) 2–4 weeks (d) Over one year

201. In the sudden wilt of guava, plants take............for complete wilting.

(a) 3 days (b) 5 days
(c) 2–4 weeks (d) Over one year

202. The predisposing factor for wilt disease of guava

(a) Stem injury (b) Root injury
(c) High temperature (d) Leaf injury

203. Guava anthracnose can be effectively managed by

(a) Spray of Copper oxychloride
(b) Application of bioagent *Streptosporangium pseudovulgare*
(c) Spray of Carbendazim or Thiophanate methyl
(d) All of these

204. Guava anthracnose is caused by

(a) *Colletotrichum psidii* (b) *Colletotrichum gloeosporioides*
(c) *Colletotrichum capsici* (d) Both (a) and (b)

205. Panama disease of banana is caused by

(a) *Fusarium oxysporum* f. sp. *cubense*
(b) *Ralstonia solanacearum* race 2
(c) *Xanthomonas campestris* pv. *musacearum*
(d) *Mycosphaerella fijiensis*

206. Moko disease of banana is caused by

(a) *Fusarium oxysporum* f. sp. *cubense*
(b) *Ralstonia solanacearum* race 2
(c) *Xanthomonas campestris* pv. *musacearum*
(d) *Mycosphaerella fijiensis*

207. Panama disease of banana produces external symptoms

(a) Green leaf syndrome (b) Yellow leaf syndrome
(c) Both (a) and (d) None of these

208. The disease cycle of the most plant pathogen can be disrupted by

(a) Crop rotation (b) Destruction of alternate host
(c) Fallow conditions (d) All of these

209. Yellow sigatoka leaf spot of banana is caused by

(a) *Mycosphaerella musicola* (b) *Mycosphaerella fijiensis*
(c) *Mycosphaerella eumusae* (d) None of these

210. Black sigatoka leaf spot of the banana is caused by

(a) *Mycosphaerella musicola* (b) *Mycosphaerella fijiensis*
(c) *Mycosphaerella eumusae* (d) None of these

211. The asexual and sexual fruiting bodies produced by the fungi causing sigatoka leaf spots of banana are

(a) Perithecia, acervuli (b) Sporodochia, perithecia
(c) Acervuli, apothecia (d) Pycnidia, perithecia

212. Banana bunchy top disease was introduced in India in

(a) 1913 (b) 1940
(c) 1928 (d) 1956

213. Banana bunchy top disease introduced in India from

(a) USA (b) Sri Lanka
(c) France (d) Germany

214. Banana bunchy top disease was introduced in Sri Lanka in

(a) 1913 (b) 1940
(c) 1928 (d) 1956

215. The characteristic severe symptoms of the banana bunchy top disease

(a) Morse-code patterns (b) Bunchy top
(c) Dark-green flecks on bract tip (d) All of these

216. *Banana bunchy top virus* particle is

(a) Isometric (b) Rod-shaped
(c) Complex (d) None of these

217. The genome of *Banana bunchy top virus* is

(a) ssDNA (b) dsDNA
(c) ssRNA (d) dsRNA

218. *Banana bunchy top virus* is primarily disseminated through

(a) Corms (b) Suckers
(c) Tissue-cultured plants (d) All of these

219. *Banana bunchy top virus* is transmitted by

(a) *Jesus indicus* (b) *Bemisia tabaci*
(c) *Pentalonia nigronervosa* (d) *Myzus persicae*

220. Papaya disease which gives a "honeycomb appearance" in tissues below the bark

(a) Foot rot (b) Anthracnose
(c) Leaf curl (d) Bunchy top

221. In papaya "shoe-string" symptoms are produced by

(a) Mosaic disease (b) Ring spot diseases
(c) Leaf curl disease (d) Anthracnose

222. Stem or foot rot of papaya is caused by

(a) *Colletotrichum gloeosporioides*
(b) *Pythium aphanidermatum*
(c) *Asperisporium caricae*
(d) *Mycosphaerella caricae*

223. "Sun up" and "Rainbow" are resistant transgenic against
(a) Mango malformation (b). Downy mildew of sunflower
(c) Papaya ring spot (d) Sigatoka leaf spot of banana

224. The most characteristic symptom of papaya leaf curl are
(a) Curling of leaves (b) Crinkling of leaves
(c) Distortion of leaves (d) All of these

225. The genome of the *Papaya leaf curl virus* is
(a) ssDNA (b) dsDNA
(c) ssRNA (d) dsRNA

226. The inoculum of *Papaya leaf curl virus* survives on
(a) Papaya (b) Tobacco and tomato
(c) Several weed hosts (d) All of these

227. *Papaya leaf curl virus* is transmitted by
(a) Aphid (b) Psyllid
(c) Leaf hopper (d) Whitefly

228. *Papaya leaf curl virus* is NOT transmitted through
(a) Sap (b) Mechanically
(c) Seed (d) All of these

229. The characteristic symptoms of papaya mosaic disese is
(a) Mild mosaic or mottling
(b) Rugose and abnormal pattern
(c) Dwarfing of plants
(d) All of these

230. Dolak Sind and Lucknow-49 are resistant cultivars against
(a) Guava wilt (b) Panama wilt of banana
(c) Anthracnose of guava (d) Apple scab

231. Domestic quarantine is operated against
(a) Wart disease of potato (b) Cyst nematode of potato
(c) Wheat rust (d) Both (a) and (b)

232. Disease cycle of the most plant pathogen can be disrupted by
(a) Crop rotation (b) Destruction of alternate/collateral hosts
(c) Fallow conditions (d) All of these

233. Little leaf of brinjal is transmitted by
(a) *Jesus indicus* (b) *Bemisia tabaci*
(c) *Pentalonia nigronervosa* (d) *Hishimonus phycitis*

234. Huanglongbing disease of citrus is transmitted by
(a) Aphid (c) Leaf hopper
(b) Psyllid (d) Whitefly

235. Which one of the following can be used to kill *Papaya mosaic potex virus*-1 and its vector?
(a) Groundnut oil (c) Paraffin oil
(b) Mustard oil (d) Kerosene oil

236. Papaya mosaic disease can be managed by the spray of
(a) 1–2% groundnut oil
(b) 5% neem oil
(c) Carbofuran 1 kg a.i. per hectare in nursery
(d) All of these

237. Papaya mosaic is NOT transmitted by
(a) Sap (b) Pruning and grafting
(c) Aphid (d) Seed

238. On papaya fruits, distinctive, small green, concentric rings, and spots, or C-shaped markings are caused by
(a) Mosaic disease (b) Ring spot disease
(c) Leaf curl disease (d) Alternaria leaf spot

239. *Papaya ring spot virus* (PRSV) is transmitted by
(a) Aphid (b) Psyllid
(c) Leaf hopper (d) Whitefly

240. *Papaya ring spot virus* (PRSV) is transmitted by
(a) *Myzus persicae* (b) *Aphis craccivora*
(c) *Aphis gossypii* (d) All of these

241. The vector of *Papaya ring spot virus* can be managed by
(a) Growing sorghum or maize as a barrier crop
(b) Use of yellow sticky trap
(c) Spry of groundnut oil
(d) All of these

242. Cross-protection involves protecting a plant from infection with
(a) Closely related to more virulent viral strain
(b) Unrelated mild viral strain
(c) Closely related mild viral strain
(d) Unrelated more virulent viral strain

243. Cross-protection is also known as
(a) Superinfection exclusion (b) Antagonism
(c) Interference (d) All of these

244. Cross-protection has not gained widespread use because
(a) Unavailability of appropriate mild strains of viruses
(b) Mild strains are not effective against all severe strains
(c) Risk of the emergence of new severe strain
(d) All of these

245. Bacterial blight of pomegranate was first reported from India by
(a) Hingorani and Mehta (1952) (b) Rane and Patel (1956)
(c) Nene and Thapliyal (1975) (d) Ramraja and Vidhyashankaran (1983)

246. Bacterial blight of pomegranate is caused by
(a) *Xanthomonas* (b) *Pseudomonas*
(c) *Ralstonia* (d) *Clavibacter*

247. L and Y shaped cracks on pomegranate fruits are caused by

(a) Fruit rot
(b) Bacterial blight
(c) Fruit cracking
(d) Fruit spot

248. The pathogen of bacterial blight of pomegranate enters and infects different plant parts through

(a) Stomata
(b) Hydathodes
(c) Wounds
(d) All of these

249. Bacterial blight of pomegranate can be managed by

(a) Spray of Copper fungicides
(b) Spray of Streptocycline + Carbendazim
(c) Spray of Streptocycline + Copper oxychloride
(d) All of these

250. The "Target board-like appearance" on the leaf is caused by

(a) *Cercospora*
(b) *Colletotrichum*
(c) *Alternaria*
(d) *Puccinia*

251. Alternaria leaf spot of *Brassica* spp. are caused by

(a) *Alternaria brassicae*
(b) *Alternaria brassicicola*
(c) *Alternaria raphani*
(d) All of these

252. Black rot of crucifers is caused by

(a) Bacteria
(b) Fungi
(c) Nematode
(d) Virus

253. Typical "V" shaped lesions on the leaf can be seen in

(a) Black spot of crucifers
(b) Black rot of crucifers
(c) Papaya ring spot
(d) Bacterial blight of pomegranate

254. Black rot of crucifers is caused by

(a) *Xanthomonas oryzae* pv. *oryzae*
(b) *Xanthomonas axonopodis* pv. *punicae*
(c) *Xanthomonas axonopodis* pv. *malvacearum*
(d) *Xanthomonas campestris* pv. *campestris*

255. Fruiting body pycnidia is produced by

(a) *Sclerotinia sclerotiorum*
(b) *Diaporthe vexan*
(c) *Macrophomina phaseolina*
(d) *Colletotrichum capsici*

256. Select the INCORRECT pair

(a) *Sclerotinia sclerotiorum*—Apothecia
(b) *Diaporthe vexan*—Pycnidia
(c) *Macrophomina phasiolina*—Cleistothecium
(d) *Colletotrichum capsici*—Acervuli

257. Phomopsis blight and fruit rot of brinjal causing fungus survives in

(a) Crop debris
(b) Seeds
(c) Soil
(d) All of these

258. The fungus *Phomopsis vexans* causes
(a) Blight and fruit rot of brinjal
(b) Damping-off of seedlings of brinjal
(c) Discolourations of brinjal seed
(d) All of these

259. Resistance/tolerant variety against Phomopsis blight and fruit rot of brinjal is
(a) Pant Brinjal Hybrid-1 (b) Pant Samrat
(c) Azad B-2 (d) All of these

260. Phomopsis blight and fruit rot of brinjal can be effectively managed by
(a) Use pathogen-free seed
(b) Hot water treatment of seed at 50°C for 30 min.
(c) Foliar spray of Carbendzim
(d) All of these

261. Diseases caused by *Sclerotinia* spp. generally known as
(a) White mold (b) Black mold
(c) Blue mold (d) Green mold

262. Chemical weapons of *Sclerotinia sclerotiorum* for attacking wide range of plants are
(a) Enzymes and oxalic acid (b) Enzymes and fumaric acid
(c) Enzymes and fussicoccin (d) Enzymes and alternaric acid

263. Which structure can be found in the pith of stems infected with *Sclerotinia*?
(a) Pycnidia (b) Sclerotia
(c) Sporodochia (d) Acervuli

264. "Damping-off" is a general term used for
(a) Death of flowers (b) Death of adult plants
(c) Death of seedlings (d) Death of branches

265. Damping off tomatoes is caused by
(a) *Phytophthora* (b) *Pythium*
(c) *Rhizoctonia* (d) All of these

266. Damping off tomatoes is can be managed by
(a) Drench the soil with Copper oxychloride
(b) Seed treatment with fungicides or bioagents
(c) Spray of Metalaxyl + Mancozeb
(d) All of these

267. Wilt pathogen colonizes
(a) Xylem (b) Phloem
(c) Cortex (d) Mesophyll

268. The wilt symptoms appear due to the blockage xylem vessel by
(a) Fungal growth and sporulation (b) Toxins, gums and gels
(c) Formation of tyloses (d) All of these

269. Tomato wilt pathogen (*Fusarium oxysporum* f. sp. *lycopersici*) disseminates through
(a) Seeds (b) Infected transplants
(c) Soil (d) All of these

270. A common sign of bacterial wilt of tomato is

(a) Spores (b) Ooze
(c) Mycelium (d) Microsclerotia

271. Green wilt in many solanaceous crops is caused by

(a) *Verticillium alboatrum* (b) *Fusarium solani*
(c) *Ralstonia solanacearum* (d) *Rhizoctonia solani*

272. *Ralstonia solanacearum* is

(a) Gram-negative, rod-shaped (b) Gram-positive, rod-shaped
(c) Gram-negative, spiral-shaped (d) Gram-negative, spherical-shaped

273. The characteristic "V-shaped" necrotic lesions on tomato leaves are produced by

(a) Verticillium wilt (b) Leaf mold
(c) Bacterial wilt (d) Early blight

274. Buckeye rot of tomato is caused by

(a) *Phytophthora nicotianae* (b) *Phytophthora infestans*
(c) *Sclerotium rolfsii* (d) *Septoria lycopersici*

275. *Phytophthora infestans* is

(a) Hemibiotrophs (b) Biotrophs
(c) Necrotrophs (d) None of these

276. Bacterial speck of tomato is caused by

(a) *Pseudomonas syringae* pv. *tomato*
(b) *Clavibacter michiganensis* subsp. *michiganensis*
(c) *Ralstonia solanacearum*
(d) *Xanthomonas campestris* pv. *campestris*

277. Bacterial canker of tomato is caused by

(a) *Pseudomonas syringae* pv. *tomato*
(b) *Clavibacter michiganensis* subsp. *michiganensis*
(c) *Ralstonia solanacearum*
(d) *Xanthomonas campestris* pv. *campestris*

278. Select the wrong pair

(a) Powdery mildew of tomato—*Leveillula taurica*
(b) Leaf mold of tomato—*Botrytis cineria*
(c) Early blight of tomato—*Alternaria linariae*
(d) Late blight—*Phytophthora infestans*

279. Leaf mold of tomato is caused by

(a) *Fulvia fulva* (b) *Botrytis cineria*
(c) *Septoria lycopersici* (d) *Alternaria solani*

280. Early blight disease of tomato and potato is also known as

(a) Alternaria leaf spot (b) Target spot
(c) Leaf spot (d) All of these

281. On tomato leaves, dark, concentric rings, resulting in a target appearance are caused by

(a) *Fulvia fulva* (b) *Botrytis cineria*
(c) *Septoria lycopersici* (d) *Alternaria solani*

282. Conditions that favours the rapid spread of early blight of tomato

(a) Warm, wet weather (b) Warm, dry weather
(c) Cool, wet weather (d) Cool, dry weather

283. The early blight fungus overwinters in

(a) Soil (b) Alternate hosts
(c) Plant debris and seed (d) All of these

284. Conidia with horizontal and vertical septa are produced by

(a) *Fulvia fulva* (b) *Botrytis cineria*
(c) *Septoria lycopersici* (d) *Alternaria solani*

285. The secondary spread of tomato early blight disease occurs by

(a) Wind and windblown soil
(b) Splashing rain
(c) Irrigation water and flea beetles
(d) All of these

286. Early blight is considered a polycyclic because of

(a) Its repeating cycles of new infection
(b) Its short disease cycle
(c) High reproductive rate of pathogen
(d) All of these

287. The effective management of late blight of tomato and potato is possible by use of

(a) Sanitation measures (b) Spray of Carbendazim + Mancozeb
(c) Spray of Ridomil (d) All of these

288. The optimum temperature for the development foot rot of papaya is

(a) 22°C (b) 28°C
(c) 32°C (d) 36°C

289. Which of the following serve as primary sources of inoculum of early blight disease of tomato?

(a) Conidia (b) Thick-walled chlamydospores
(c) Dark pigmentation mycelia (d) All of these

290. Polycyclic pathogens are disseminated primarily by

(a) Irrigation water (b) Air or airborne vectors
(c) Seed (d) Soil

291. Examples of polycyclic diseases are

(a) Downy mildews and powdery mildews
(b) Leaf spots and blights
(c) Grain rusts and insect-borne viruses
(d) All of these

292. In polycyclic fungal pathogens, the primary inoculum often consists of

(a) Sexual spores (b) Sclerotia
(c) Mycelium in infected tissue (d) All of these

293. Buckeye rot of tomato is caused by
(a) Nutrient deficiency (b) A virus
(c) A fungus (d) An Oomycetes

294. Non-parasitic disorder due to deficiency of calcium in tomato is
(a) Cat face (b) Leaf curling
(c) Buttoning (d) Blossom end rot

295. The cat-face is a physiological disorder of
(a) Potato (b) Tomato
(c) Capsicum (d) Pea

296. The optimum temperature at which sporangia of *Phytophthora infestans* produce zoospores
(a) 16°C (b) 12°C
(c) 21°C (d) 25°C

297. Buckeye rot is a disease of
(a) Okra (b) Tomato
(c) Brinjal (d) Apple

298. Buckeye rot of tomato can be diagnose by
(a) Bull's eye pattern on fruits
(b) Fruit lesions that remains firm and smooth
(c) Fruit lesions are with water-soaked zonations
(d) All of these

299. Tomato fruit with latent infections of buckeye rot may decay during transit and storage if the temperature is
(a) 10°C (b) 12°C
(c) 15°C (d) 21°C

300. The buckeye rot pathogen is
(a) Externally seedborne (b) Internally seedborne
(c) Airborne only (d) Soilborne only

301. The buckeye rot pathogen of tomato may be introduced through
(a) Contact with infested soil
(b) Infected seeds or transplants
(c) Plants from the previous crop
(d) All of these

302. Tomato yellow leaf curl is also known as
(a) Curly top disease (b) Green wilt disease
(c) Cat face disease (d) Yellow mosaic disease

303. Symptoms of tomato yellow leaf curl disease on the leaf include
(a) Mosaic and vein clearing (b) Crinkling and puckering
(c) Inward rolling and chlorosis (d) All of these

304. ***Tomato yellow leaf curl virus* is**

(a) Caulimovirus (b) Potyvirus
(c) Begomovirus (d) Comovirus

305. ***Tomato yellow leaf curl virus* (TYLCV) is transmitted by**

(a) Whitefly (b) Aphid
(c) Thrips (d) Plant hopper

306. Buckeye rot of tomato is also known as

(a) Soft rot (b) Phytophthora root rot
(c) Fruit rot (d) All of these

307. Buckeye rot incidence may reach upto 90% under

(a) High humidity and good rainfall
(b) Wet and cool condition
(c) Dry and cool condition
(d) High temperature and stress condition

308. Tomato yellow leaf curl disease can be managed by

(a) Using virus-free transplants
(b) Removing the virus-affected plants
(c) Installing yellow sticky traps
(d) All of these

309. Resistant variety of tomato to yellow leaf curl disease is

(a) Kashi Vishesh (b) Arka Ananya
(c) Kashi Amrit (d) All of these

310. The best long-term practice to manage plant diseases is

(a) Use of chemicals (b) Use of resistant varieties
(c) Use of biopesticides (d) Use of pathogen-free seeds

311. Which of the following viruses are highly infectious, extremely persistent, and easily transmits during handling of tomato plants?

(a) *Tobacco mosaic virus* and *Tomato mosaic virus*
(b) *Tobacco mosaic virus* and *Cucumber mosaic virus*
(c) *Potato virus X* and *Potato virus Y*
(d) None of these

312. A very damaging tomato syndrome called "double streak" is caused by a co-infection of

(a) *Tobacco mosaic virus* and *Tomato mosaic virus*
(b) *Tobacco mosaic virus* and *Cucumber mosaic virus*
(c) *Potato virus X* and *Potato virus Y*
(d) *Tobacco mosaic virus* and *Potato virus X*

313. Most characteristic feature of *Cucumber mosaic virus* (CMV) infection in tomato is

(a) Shoestring of leaflets (b) Curling of leaflets
(c) Crinkling of leaflets (d) Puckering of leaflets

314. Virus-induced extremely filiform leaves is called
(a) Shoelace (b) Shoestring
(c) Fernleaf (d) All of these

315. The virus with ssRNA genome is
(a) *Tobacco mosaic virus*. (b) *Potato virus Y*
(c) *Cucumber mosaic virus* (d) All of these

316. TMV and ToMV can survive in plant debris for upto
(a) 5 years (b) 15 years
(c) 50 years (d) 100 years

317. The rate of seed to seedlings transmission of *Tomato mosaic virus* (ToMV) is
(a) 1%–10% (b) 1%–13%
(c) 10%–20% (d) 20%–30%

318. The rate of seed to seedlings transmission of *Tobacco mosaic virus* (TMV) is
(a) 1%–10% (b) 1%–13%
(c) 10%–20% (d) 20%–30%

319. Plant viruses which are NOT transmitted by insects
(a) *Tobacco mosaic virus* and *Tomato mosaic virus*
(b) *Tomato leaf curl virus* and *Banana bunchy top virus*
(c) *Potato leaf roll virus* and *Potato virus Y*
(d) *Papaya ring spot virus* and *Okra yellow vein mosaic virus*

320. Geminivirus has
(a) ssDNA molecule (b) Twin genome
(c) Betasatellite (d) All of these

321. Resistant or tolerant variety to yellow vein mosaic of okra
(a) Arka Anamika (b) Parbhani Kranti
(c) Hisar Naveen (d) All of these

322. Yellowing of veins and veinlets of okra is caused by
(a) Nematode (b) Virus
(c) Fungus (d) Bacteria

323. Yellow vein mosaic of okra can be managed by
(a) Rouging of infected plants
(b) Installing yellow sticky traps
(c) Spray of Chlorpyriphos + Neem oil
(d) All of these

324. For the management of yellow vein mosaic of okra which border crop can be used to reduce whitefly infestations?
(a) Maize (b) Sorghum
(c) Pearl millet (d) All of these

325. Anthracnose of beans is caused by
(a) *Colletotrichum gloeosporioides* (b) *Colletotrichum lindemuthianum*
(c) *Colletotrichum acutatum* (d) *Colletotrichum falcatum*

326. Select the wrong pair
- (a) Sunken dead spots—Anthracnose
- (b) Tissue browning—Wilt
- (c) Dusty mass of spores on the host—Smut
- (d) Cell death—Necrosis

327. Bean anthracnose is favoured by
- (a) Heavy and frequent rains
- (b) Cool and wet weather
- (c) Temperature range (13°C–26°C)
- (d) All of these

328. Common blight of bean is caused by
- (a) *Xanthomonas axonopodis* pv. *phaseoli*
- (b) *Pseudomons syringae* pv. *phaseolicola*
- (c) *Xanthomonas phaseoli* pv. *sojensis*
- (d) *Pseudomons tabaci*

329. In the disease cycle of bean anthracnose, the pathogen first switch to
- (a) Necrotrophic lifestyle
- (b) Saprophytic lifestyle
- (c) Necrotrophic then biotrophic lifestyle
- (d) Biotrophic then necrotrophic lifestyle

330. Select the wrong pair
- (a) Oxycarboxin—Vitavex
- (b) Carboxin—Plantvex
- (c) Copper oxychloride—Blitox-50
- (d) Tricyclazole—Tilt

331. Which of the following is a seedborne disease?
- (a) Bean anthracnose
- (b) Early blight of tomato
- (c) Phomopsis blight of brinjal
- (d) All of these

332. Seedborne inoculum of bean anthracnose can be killed by seed treatment with
- (a) Carbendazim
- (b) Benlate
- (c) Mancozeb
- (d) All of these

333. Common and fuscous blight of bean is caused by
- (a) Fungus
- (b) Bacteria
- (c) Virus
- (d) Phytoplasma

334. The dissemination of common blight of beans occurs through
- (a) Wind-splashed rain
- (b) Infected seed
- (c) Insects
- (d) All of these

335. Halo blight of beans is caused by
- (a) *Xanthomonas axonopodis* pv. *phaseoli*
- (b) *Pseudomons syringae* pv. *phaseolicola*
- (c) *Colletotrichum lindemuthianum*
- (d) *Pseudomons tabaci*

336. The source of secondary infections of halo blight of beans is
(a) Seedborne inoculum (b) Bacterial ooze from seedling lesions
(c) Infected crop residue (d) Infected plant debris on the soil

337. The halo blight disease of bean is favoured by
(a) 10°C–15°C (b) 16°C–23°C
(c) 20°C–25°C (d) 25°C–30°C

338. The development of yellow haloes and systemic chlorosis in bean is inhibited
(a) At 18°C (b) At 20°C
(c) Below 27°C (d) Above 27°C

339. The bacterial toxin responsible for systemic chlorosis in halo blight of beans
(a) Tabtoxin (b) Syringomycin
(c) Phaseolotoxin (d) Coronotine

340. Pythium soft rot of ginger was first time recorded in India by
(a) R.S. Singh (1985) (b) E.J. Butler (1907)
(c) H.S. Chaube (1983) (d) P.R. Verma (1976)

341. Colocasia leaf blight is caused by
(a) *Alternaria solani* (b) *Phytophthora colocasiae*
(c) *Botrytis cineria* (d) *Phytophthora nicotianae*

342. Select the wrong pair
(a) Carbendazim—Bavistin
(b) Mancozeb—Dithane M45
(c) Metalaxyl + Mancozeb—Ridomil MZ
(d) Carbendazim + Mancozeb—Curzate M8

343. Wrong statement about *Phytophthora colocasiae*
(a) It is an Oomycetes (b) It produces septate hyphae
(c) It produces oospores (d) It produces long, unbranched haustoria

344. *Phytophthora colocasiae* (taro leaf blight) survive
(a) In infected crop residues
(b) As mycelium and encysted zoospores
(c) As chlamydospores and oospores
(d) All of these

345. Phytophthora leaf blight of Colocasia is favoured by
(a) Long period of leaf wetness, high humidity (100%), temperature (20°C–30°C)
(b) Short period of leaf wetness, high humidity (100%), temperature (15°C–20°C)
(c) Long period of leaf wetness, low humidity (50%), temperature (30°C–35°C)
(d) Short period of leaf wetness, low humidity (70%), temperature (10°C–20°C)

346. Most commonly recommended fungicide taro leaf blight (*Phytophthora colocasiae*) is
(a) Mancozeb (b) Ridomil MZ72
(c) Phosphorus acid (d) All of these

347. Bud rot of coconuts was reported in India by
(a) E.J. Butler (1906) (b) G.M. Das (1965)
(c) C. Mohanan (1978) (d) G. Rangaswami (1979)

348. Bud rot of coconuts is caused by
(a) *Ganoderma lucidum* (b) *Phytophthora nicotianae*
(c) *Phytophthora palmivora* (d) *Ceratostomella paradoxa*

349. Stem bleeding of coconuts is caused by
(a) *Ganoderma lucidum* (b) *Phytophthora nicotianae*
(c) *Phytophthora palmivora* (d) *Ceratostomella paradoxa*

350. Basal stem end rot of coconuts is caused by
(a) *Ganoderma lucidum* (b) *Phytophthora nicotianae*
(c) *Phytophthora palmivora* (d) *Ceratostomella paradoxa*

351. Tanjore wilt of coconuts is also known as
(a) Root rot or wilt (b) Basal stem end rot
(c) Ganoderma wilt (d) All of these

352. Basal stem end rot of coconut can be very effectively managed by intercropping with
(a) Banana (b) Hemp
(c) Turmeric (d) All of these

353. Blister blight of tea is caused by
(a) *Pestalotia theae* (b) *Exobasidium vexans*
(c) *Pellicularia koleroga* (d) *Cephaleuros mycoids*

354. *Exobasidium vexans* is
(a) A bacteria (b) A fungus
(c) An insect (d) A nematode

355. *Exobasidium vexans* is
(a) Facultative parasite (b) An obligate parasite
(c) Facultative saprophyte (d) None of these

356. *Exobasidium vexans* is disseminated through
(a) Airborne ascospores (b) Airborne basidiospores
(c) Soilborne oospores (d) Soilborne zygospores

357. Infection-related structure of plant pathogenic fungi is
(a) Acervuli (b) Appressoria
(c) Inoculum (d) Sclerotia

358. Which of the following is NOT a polycyclic disease?
(a) Early blight of tomato (b) Blister blight of tea
(c) Loose smut of wheat (d) Late blight of potato

359. The life cycle of *Exobasidium vexans* (blister blight of tea) completes in
(a) 5–8 days (b) 10–15 days
(c) 11–28 days (d) 25–35 days

360. Which of the following can manage the blister blight of tea effectively?
(a) Hexaconazole (b) *Orchobacterium anthropi* BMO-111
(c) Copper oxychloride (d) Mancozeb

361. Leaf rust of coffee introduced in India from
(a) USA (b) Sri Lanka
(c) France (d) Bangladesh

362. Leaf rust of coffee introduced in India in
(a) 1888 (b) 1879
(c) 1942 (d) 1901

363. Leaf rust of coffee is caused by
(a) *Hemileia vastatrix* (b) *Exobasidium vexans*
(c) *Puccinia triticina* (d) *Uromyces fabae*

364. *Hemileia vastatrix* is
(a) A biotrophic and microcyclic (b) A necrotrophic and demicyclic
(c) A biotrophic and macrocyclic (d) A hemibiotrophic and macrocyclic

365. The generic name of the fungus *Hemileia* meaning
(a) Half-life (b) Half-smooth
(c) Oval shaped (d) Plant destroyer

366. The uredospores of *Hemileia vastatrix* are
(a) Kidney-shaped (b) Pear shaped
(c) Oval shaped (d) Filiform

367. Hyperparasite which can be used as bioagent against coffee rust fungus
(a) *Verticillium hemileiae* (b) *Paranectris hemileiae*
(c) *Darluca filum* (d) All of these

368. Fungicide formulation with systemic and contact mode of action
(a) Tilt (b) Bavistin
(c) Ridomil MZ (d) Blitox-50

369. 'Green islands' formations in leaves of many plants are typical infections of
(a) Biotrophic pathogens (b) Necrotrophic pathogens
(c) Saprphytic parasites (d) Facultative parasites

370. 'Green islands' can be seen in
(a) Powdery mildew and rust diseases
(b) Wilt and blight diseases
(c) Leaf spots and smut diseases
(d) Sooty mold and red rust diseases

371. The changes in plant development can be caused by a biotrophic pathogen
(a) Hypertrophy
(b) Hyperplasia
(c) Abnormal differentiation of organs
(d) All of these

372. Heart rot of beet is caused by the deficiency of
(a) Zinc (b) Molybdenum
(c) Calcium (d) Boron

373. Green ear disease of bajara is an example of
(a) Endemic disease
(b) Sporodic disease
(c) Epidemic disease
(d) Pandemic disease

374. The initial stage of ergot infection is known as
(a) Honeydew stage
(b) Sclerotial stage
(c) Spore formation stage
(d) None of these

375. Sclerotia of ergot can be removed from the seed by floating them on
(a) Glycerol
(b) Salt solution
(c) Mustard oil
(d) Kerosene oil

376. *Papaya leaf curl virus* causes
(a) Severe curling on leaves
(b) Crinkling and distortion of leaves
(c) Vein clearing and reduction of leaf size
(d) All of these

377. Whitefly transmits
(a) *Papaya leaf curl virus*
(b) *Tomato leaf curl virus*
(c) *Okra yellow vein mosaic virus*
(d) All of these

378. Rice plants show markedly stunted leaves, yellow to orange discolouration and interveinal chlorosis. This may be due to
(a) Zinc deficiency
(b) *Rice tungro virus*
(c) *Rice dwarf virus*
(d) *Xanthomonas oryzae* pv. *oryzae*

379. Alternate host for the rust of pearl millet
(a) Maize
(b) Brinjal
(c) Tomato
(d) Okra

380. Moko disease of banana is caused by
(a) *Ralstonia solanacearum*
(b) *Fusarium oxysporum* f.sp. *cubense*
(c) *Mycospharella musicola*
(d) *Verticillium dahlia*

381. Panama disease is a devastating disease of
(a) Mango
(b) Apple
(c) Banana
(d) Sugarcane

382. *Hemelia vastatrix* causes
(a) Bean rust
(b) Coffee leaf rust
(c) Yellow rust of wheat
(d) Sunflower rust

383. Bunchy top of banana is transmitted by
(a) *Bemisia tabaci*
(b) *Pentalonia nigranervosa*
(c) *Myzus persicae*
(d) *Aphis gossypii*

384. Sorghum smut which is NOT a seedborne disease
(a) Loose smut
(b) Grain smut
(c) Head smut
(d) Long smut

385. Translucent water-soaked lesion with wavy edges and yellow margins near the tip is the characteristic symptoms of
(a) Bacterial blight of cotton (b) Bacterial blight of rice
(c) Red rot of sugarcane (d) Stalk rot of maize

386. Halo blight of beans is caused by
(a) *Xanthomonas campestris* pv. *phaseolicola*
(b) *Pseudomonas syringae* pv. *phaseolicola*
(c) *Pseudomonas syringae* pv. *syringae*
(d) *Xanthomonas oryzae* pv. *oryzae*

387. The mode of infection in ergot of pearl millet is
(a) Leaf infection (b) Stem infection
(c) Blossom infection (d) Root infection

388. Symptoms produced by *Sclerospora graminicola* in pearlmillet mainly comprise of
(a) Shredding of leaves (b) Downy mycelial growth
(c) Green ears (d) All of these

389. Which of the following is required during the infection process of *Magnaporthe oryzae*?
(a) Melanin (b) penetration peg
(c) Turgor pressure (d) All of these

390. Reniform uredospores are produced by
(a) *Hemileia* (b) *Puccinia*
(c) *Melampsora* (d) *Uromyces*

391. In *Claviceps*, part spores form conidia rather than germ tubes and the process is called
(a) Iterative germination (b) Microcyclic conidiation
(c) Direct germination (d) Both (a) and (b)

392. Disease-related to groundnut is
(a) Red rot (b) Tikka
(c) Black arm (d) Loose smut

393. "Buckeye rot" is a disease of
(a) Sweet potato (b) Tomato
(c) Rice (d) Garden pea

394. The plant pathogen in which the thallus is both epibiotic as well as endobiotic
(a) *Physoderma* (b) *Olpidium*
(c) *Synchytrium* (d) All of these

395. Mostly downy mildew of cereals is caused by
(a) *Sclerospora* sp. (b) *Peronospora sp.*
(c) *Plasmopara* sp. (d) *Pseudoperonospora sp.*

396. Spray with 0.1% groundnut oil cake help to prevent the infection of
(a) Papaya mosaic (b) Leaf roll of potato
(c) Pigeon pea sterility mosaic (d) Witche's broom of potato

397. Soybean yellow mosaic transmitted by

(a) Whitefly (b) Seed
(c) Leaf hopper (d) Aphid

398. Blossom end rot of capsicum and tomato is caused by the deficiency of

(a) Boron (b) Calcium
(c) Zinc (d) Molybdenum

399. The alternate host of pearl millet rust is

(a) Maize (b) Brinjal
(c) Sorghum (d) Radish

400. Stem bleeding of coconut is caused by

(a) *Phytophthora palmivora* (b) *Ceratostomella paradoxa*
(c) *Fomes annosus* (d) *Rhizoctonia solani*

401. Pale yellow ooze from the cut surfaces of banana appears in

(a) Panama wilt (b) Banana Xanthomonas wilt
(c) Moko disease (d) Sigatoka leaf spot

402. The optimum temperature range for the production of phaseolotoxin by *Pseudomonas syringae* pv. *phaseolicola* is

(a) 2°C–8°C (b) 18°C–20°C
(c) 7°C–10°C (d) 27°C–30°C

403. Bacterial blights of common beans can be managed mainly by

(a) Sowing of pathogen-free seed (b) Growing resistant varieties
(c) Crop rotation (d) All of these

404. The disease name halo blight came from the characteristic yellow halo that surrounds water-soaked lesions on a bean leaf infected with *P. syringae* pv. *phaseolicola*. The chlorotic or yellow halo results from the action of

(a) Pectolytic enzymes (b) Siderophore of pathogen
(c) Nonspecific phaseolotoxin (d) Water-soluble pigment of pathogen

405. Tikka disease of groundnut is caused by

(a) *Cercosporidium personatum* (b) *Cercospora arachidicola*
(c) *Cercospora musicola* (d) Both (a) and (b)

406. In groundnut........................cause "Early leaf spot" while..................causes "Late leaf spot" caused by................

(a) *Cercospora arachidicola*, *Cercospora personata*
(b) *Cercospora personata*, *Cercospora arachidicola*
(c) *Cercospora musicola*, *Cercospora capsici*
(d) *Alternaria solani*, *Phytophthora infestans*

407. Which of thc following fungicide is used for seed treatment?

(a) Mancozeb (b) Diathane
(c) Thiram (d) All of these

408. Carbendazim is

(a) Fungicide (b) Insecticide
(c) Rodenticide (d) Nematicides

409. The correct sequence of seed treatment with fungicide, rhizobium/bioagents, and insecticide is

(a) Fungicide, rhizobium/bioagents, insecticide
(b) Rhizobium/bioagents, fungicide, insecticide
(c) Fungicide, insecticide, rhizobium/bioagents
(d) Insecticide, rhizobium/bioagents, fungicide

410. Grain smut of sorghum is also known as

(a) Kernel smut (b) Covered smut
(c) Short smut (d) All of these

411. Loose smut of sorghum is caused by

(a) *Ustilago tritici* (b) *Tolyposporium ehrenbergii*
(c) *Ustilago maydis* (d) *Sphacelotheca cruenta*

412. Ergot disease of bajra is caused by

(a) *Sclerospora graminicola* (b) *Colletotrichum graminicola*
(c) *Ustilago maydis* (d) *Claviceps fusiformis*

413. In pearl millet, the deformity is characterised by the complete transformation of floral organs into the twisted leafy structures. This gives ear an appearance of the green leafy mass hence named

(a) Green ear (b) Phyllody
(c) Leaf roll (d) Witche's broom

414. In maize "Crazy Top" symptoms are the characteristics of

(a) Sheath blight (b) Common smut
(c) Stalk rot (d) Downy mildew

415. Phomopsis blight is severe in

(a) High RH and high temperature
(b) Low RH and low temperature
(c) Low RH and high temperature
(d) None of these

416. Seedborne inoculum of Phomopsis blight can be reduced by

(a) Seed treatment with Carbendazim
(b) Seed treatment with Captan
(c) Seed treatment with Mancozeb
(d) Seed treatment with Copper oxychloride

417. Perfect stage of *Phomopsis vexans* is

(a) *Diplocarpon rosae* (b) *Diaporthe vexans*
(c) *Phoma* spp. (d) *Phyllosticta* spp.

418. *Phomopsis vexans* overwinters in

(a) Seed (b) Plant debris
(c) Seed and plant debris (d) None of these

419. *Sclerotinia sclerotiorum* overwinters in the form of

(a) Apothecia (b) Sclerotia
(c) Perithecia (d) Cleistothecia

420. Sclerotia of *Sclerotinia sclerotiorum* germinates by producing
(a) Apothecium (b) Conidiomata
(c) Pycnidium (d) Basidium

421. Optimum temperature for *Sclerotinia sclerotiorum* infection is
(a) 0°C–10°C (b) 10°C–15°C
(c) 15°C–20°C (d) 25°C–30°C

422. Cercospora leaf spots are severe during
(a) Warm days and cool nights (b) Cool days and cool nights
(c) Warm days and warm nights (d) None of these

423. Local transmission of *Cercospora* spp. is helped by
(a) Moist wind and irrigation water
(b) Dry wind and irrigation water
(c) Moist wind and rainfed conditions
(d) None of these

424. Incidence of bacterial wilt increase with
(a) Decrease in altitude (b) Increase in altitude
(c) On intermediate elevations (d) None of these

425. Little leaf disease is more serious in
(a) Southeast Asian countries (b) South African countries
(c) Southwest Asian countries (d) None of these

426. Little leaf disease is caused by
(a) Fungi (b) Bacteria
(c) Virus (d) Phytoplasma

427. Little leaf symptoms are characterised by
(a) Reduction in leaf size having shortened petioles
(b) Reduction in branches
(c) Reduction in size of stems
(d) None of these

428. Little leaf is transmitted by
(a) Leaf hopper (b) Aphids
(c) Whitefly (d) Thrips

429. Littlc lcaf Phytoplasma perennates in
(a) Soil (b) Weed hosts
(c) Seed (d) None of these

430. Alpha spores of *Phomposis vexans* are converted into beta spores at
(a) 10°C–16°C (b) 20°C–25°C
(c) 25°C–30°C (d) 25°C–35°C

431. Tomato leaf curl disease is transmitted by
(a) Thrips (b) Whitefly
(c) Aphids (d) Leaf hoppers

432. Leaf curl of tomato is a geminivirus whose nucleic acid is made up of

(a) RNA (b) DNA
(c) RNA and DNA (d) ssRNA

433. *Papaya ring spot virus* is a nonpersistent virus transmitted by

(a) *Toxoptera citricidus* (b) *Myzus persicae*
(c) *Bemisia tabaci* (d) *Cuscuta reflexa*

434. Powdery mildew of chilli is caused by

(a) *Erysiphe polygoni* (b) *Erysiphe cichoracearum*
(c) *Leveillula taurica* (d) *Sphaerotheca fuliginia*

435. Highly contagious virus

(a) *Cucumber mosaic virus* (b) *Tobacco mosaic virus*
(c) *Potato leaf roll virus* (d) *Tomato leaf roll virus*

436. *Exobasidium vexans* which causes

(a) Blight of coffee (b) Blister blight of tea
(c) Blister blight of cocoa (d) Red rust of tea

437. Early blight of potato and tomato caused by

(a) *Alternaria solani* (b) *Alternaria alternata*
(c) *Alternaria helianthi* (d) *Alternaria tenuis*

438. Akiochi disease of rice is due to

(a) Hydrogen sulfide toxicity (b) Chloride toxicity
(c) Zinc deficiency (d) Manganese deficiency

439. Rice variety Ajaya is resistant to

(a) Bacterial blight (b) Gall midge
(c) Zinc deficiency (d) Tungro

440. Rice variety Vikramaraya is resistant to

(a) Bacterial blight (b) Gall midge
(c) Zinc deficiency (d) Tungro

441. The pathogens causing rust disease in crops are

(a) Obligate saprophytes (b) Facultative saprophyte
(c) Facultative parasite (d) Obligate parasite

442. Green ear disease of pearl millet is caused by

(a) *Tolyposporium penicillariae* (b) *Sclerospora graminicola*
(c) *Claviceps microcephala* (d) *Curvularia penniseti*

443. The *mungbean yellow mosaic virus* is transmitted by

(a) Whiteflies (b) Sap
(c) Seeds (d) All of these

444. An example of gram-positive bacteria is

(a) *Erwinia* (b) *Pseudomonas*
(c) *Xanthomonas* (d) *Clavibactor*

445. Pathogen that exploit the host tissue after killing it with toxins and enzymes
(a) *Septoria* (b) *Fusarium*
(c) *Alternaria* (d) All of these

446. Which one of the following pathogens does NOT cause wilt disease?
(a) *Fusarium* (b) *Verticillium*
(c) *Meloidogyne* (d) *Erwinia*

447. Which one of the following symptoms can be associated with viral, fungal, or bacterial infections as well as insect and mite infestations?
(a) Root rot (b) Leaf spot
(c) Wilting (d) Leaf distortion

448. The term which describes the rapid death and collapse of young seedlings is called
(a) Stem rot (b) Root proliferation
(c) Wilting (d) Damping-off

449. Damping-off disease in the plant is caused by
(a) *Pythium* (b) *Rhizoctonia*
(c) *Thielaviopsis* (d) All of these

450. Which one of the following pathogens does NOT require a living host in order to grow?
(a) Powdery mildew fungi (b) Sclerotia-producing fungi
(c) Phytoplasma (d) Viruses

451. Geminiviruses are mostly transmitted by
(a) Aphids (b) Whitefly
(c) Leaf hoppers (d) Thrips

452. Sterility mosaic disease of pigeon pea is transmitted by
(a) Eriophyid mite (b) Jassids
(c) Whitefly (d) Aphids

453. The vector involves in the transmission of phyllody disease of sesamum is
(a) *Nephotettix virescens* (b) *Myzus persicae*
(c) *Urocystis brassicae* (d) *Orosius albicinctus*

454. Rust of flax is caused by
(a) *Puccinia graminis* (b) *Melampsora lini*
(c) *Uromyces pisi* (d) *Uromyces appendiculatus*

455. Mycorrhiza helps in the uptake of
(a) Nitrate (b) Potassium
(c) Phosphorus (d) Boron

456. Black arm disease of cotton is caused by
(a) *Xanthomonas campestris* pv. *citri*
(b) *Ralstonia solanacearum*
(c) *Xanthomonas campestris* pv. *malvacearum*
(c) *Pseudomonas syringae*

457. "Phyllody disease" of sesamum is spread by
(a) Leaf hopper (b) Stem borer
(c) Mite (d) White fly

458. Bunchy top of papaya is transmitted by
(a) Whitefly (b) Stem borer
(c) Aphids (d) Leaf hopper

459. The pathogen responsible for causing "charcoal rot of soybean" is
(a) *Macrophomina phaseolina* (b) *Glomerella lindemuthianum*
(c) *Uromyces phaseoli* (d) *Erysiphe polygoni*

460. The causal agent of rice bunt disease
(a) *Tilletia barclayana* (b) *Alternartia padwiki*
(c) *Pyricularia oryzae* (d) *Rhizoctonia solani*

461. The causal agent of "udbatta disease" of rice
(a) *Fusarium fujikuroi* (b) *Bipolaris oryzae*
(c) *Ephelis oryzae* (d) *Pyriculria oryzae*

462. A localised area of tissue, usually of a leaf, which remains green around a pathogenic infection and against a background or general spreading chlorosis is referred as
(a) Smut (b) Green islands
(c) Powdery mildew (d) Downy mildew

463. Fungi produce sclerotia for
(a) Rapid multiplications
(b) Surviving under unfavourable conditions
(c) Infecting the host plants
(d) Developing apothecia

464. Abnormal development of the most of plant parts causing the reduction in plant height, stem length, leaf size, and other organs is called
(a) Smut (b) Downy mildew
(c) Dwarfing (d) Enation

465. The "panama wilt" disease can be observed in
(a) Mango (b) Banana
(c) Apple (d) Pomegranate

466. Disease caused in man and other animals by the toxic substances in the ergots, which contaminate cereal grains, is referred as
(a) Ergotism (b) Frenching
(c) Mycosis (d) Botulism

467. Viruses produce symptoms similar to those of
(a) Fungal infection (b) Bacterial infection
(c) Nutritional deficiency (d) None of these

468. Dieback disease is caused by
(a) Magnesium deficiency (b) Zinc deficiency
(c) Fungi (d) Virus

469. Sheath rot of rice is caused by
(a) *Rhizoctonia solani* (b) *Ephelis oryzae*
(c) *Sclerotium rolfsii* (d) *Sarocladium oryzae*

470. Which of the following plant virus has ssDNA genome?
(a) *Cauliflower mosaic virus* (b) *Banana bunchy top virus*
(c) *Rice dwarf virus* (d) *Rice tungro bacilliform virus*

471. Soybean rust is caused by
(a) *Puccinia gramins tritici* (b) *Phakopsora pachyrhizi*
(c) *Uromyces appendiculatus* (d) *Uromyces fabae*

472. "Akiochi" disease in rice is due to the toxicity of
(a) Hydrogen citrate (b) Hydrogen chloride
(c) Hydrogen peroxide (d) Hydrogen sulphide

473. The term "Anthracnose" means
(a) Coal-like disease (b) Ulcer-like lesions
(c) Black spot (d) All of these

474. Which of the following is NOT a physiological disease?
(a) Black heart of potato (b) Tip burn of paddy
(c) Dieback of linseed (d) Marsh spot of pea

475. Which mycotoxin is produced by *Aspergillus flavus* strains on groundnut and cereals?
(a) Mycotoxin (b) Aflatoxin
(c) Cytotoxin (d) Endotoxin

476. Big bud of tomato is caused by
(a) Fungi (b) Nematode
(c) Virus (d) Phytoplasma

477. Leaf curl of tobacco is caused by
(a) Fungi (b) Bacteria
(c) Nematode (d) Virus

478. Symptoms produced by the plant viruses
(a) Mosaics (b) Enation
(c) Necrosis (d) All of these

479. Which of the following is NOT an asexual fruiting body?
(a) Sporangium (b) Apothecia
(c) Conidiomata (d) Zoosporangium

480. A blister-like symptom that may cause the epidermis to rupture, revealing the spores of the pathogen
(a) Spot (b) Necrosis
(c) Blight (d) Pustules

481. Toxin pyricularin is produced by
(a) Fungus (b) Bacteria
(c) Virus (d) Nematode

482. Bhindi yellow vein clearing disease is transmitted by
(a) *Aphis gossypii* (b) *Oligonychus indicus*
(c) *Bemisia tabaci* (d) *Amrasca devastans*

483. Biotrophic pathogens derive nutrients from
(a) Dead cells (b) Damaged cells
(c) Necrotic cells (d) Living cells

484. Which of the following is a sign of bacterial plant disease?
(a) Bacterial ooze
(b) Water-soaked lesions
(c) Bacterial streaming in water from a cut stem
(d) All of these

485. Which one of the following is NOT a soil-borne plant pathogens?
(a) *Gaeumannomyces graminis* (b) *Rhizoctonia solani*
(c) *Armillaria mellea* (d) *Pyricularia oryzae*

486. The least enzymatic activity is seen with
(a) Necrotrophic pathogen (b) Biotrophic pathogen
(c) Hemibiotroph pathogen (d) All of these

487. Which one of the following is NOT a viral disease symptom?
(a) Mosaic leaf pattern (b) Crinkled leaves
(c) Plant stunting (d) Root galls

488. The loss of turgidity of leaves is called
(a) Wilting (b) Leaf curling
(c) Necrosis (d) Leaf rolling

489. Leaf wilting is a typical symptom of
(a) Verticilium wilt (b) Fusarium wilt
(c) Bacterial wilt (d) Physiological wilt

490. A mosaic symptom refers to the
(a) Light and dark green patterns on the leaves
(b) Yellowing of the leaves
(c) Greening of the petals
(d) Excessive proliferation of the roots

491. The diseases which affect the entire plant
(a) Wilts and diebacks
(b) Smuts and rusts
(c) Leaf spots and leaf curls
(d) Canker and gummosis

492. In which of the following cases, the overdevelopment of tissues or organs occurs?
(a) Galls on roots or leaves (b) Witches' brooms
(c) Profuse flowering (d) All of these

493. Target spot lesion on tobacco is caused by
(a) *Alternaria solani* (b) *Phyllosticta nicotiana*
(c) *Ascochyta phaseolorum* (d) *Rhizoctonia solani*

494. Which of the following microorganism causes fatal poisoning in canned fruits and vegetables?
(a) *Aspergillus flavus* (b) *Penicillium digitatum*
(c) *Clostridium botulinum* (d) *Rhizoctonia solani*

495. Fusarium wilt of the tomato is
(a) Seedborne (b) Soilborne
(c) Airborne (d) All of these

496. *Ralstonia solanacearum* causes wilt in
(a) Tomato (b) Banana
(c) Ginger (d) All of these

497. Sigatoka disease of banana is caused by
(a) *Cercospora personatum* (b) *Cercosopra musae*
(c) *Alternaria padwickii* (d) *Albugo candida*

498. The causal organism of "potato blackleg" is
(a) *Erwinia amylovora*
(b) *Ralstonia solanacearum*
(c) *Pectobacterium carotovora*
(d) None of these

499. Genetic material of *Tobacco mosaic virus* is
(a) DNA (b) RNA
(c) DNA or RNA (d) DNA and RNA

500. Green ear disease of pearl millet is caused by
(a) *Pernosclerospora sorghi* (b) *Hyaloperonospora parasitica*
(c) *Plasmopara viticola* (d) *Sclerospora graminicola*

501. Often......helps in infection by *Fusarium*, hence these pathogens should be managed.
(a) Bacteria (b) *Ralstonia solanacearum*
(c) Phytoplasma (d) Root knot nematode

502. Papaya bunchy top is caused by
(a) Fungi (b) Bacteria
(c) Phytoplasma (d) Virus

503. Vegetative spore of *Fusarium oxysporum* is
(a) Oospore (b) Ascopspore
(c) Chlamydospore (d) Basidiospore

504. Systemic fungicide used for the management of powdery mildew disease is
(a) Sulphur dust (b) Carboxin
(c) Calixin (d) Carbendazim

505. *Ralstonia solanacearum* is
(a) Gram-positive, rod shaped
(b) Gram-negative, rod shaped
(c) Gram-positive, spherical shaped
(d) Gram-negative, spherical shaped

506. India's first triple disease-resistant tomato F1 hybrid "Arka Rakshak" is resistant to
(a) Late blight, bacterial canker, and root knot nematode
(b) Late blight, fusarium wilt, and root knot nematode
(c) Buckeye rot, bacterial wilt, and tomato mosaic
(d) Early blight, bacterial wilt, and *Tomato yellow leaf curl virus*

27.2 MATCH THE FOLLOWING

1. (i) False smut of rice — (A) *Pyricularia oryzae*
(ii) Kernel smut of rice — (B) *Fusarium fujikuroi*
(iii) False ergot of rice — (C) *Neovossia barclayana*
(iv) Bakanae disease of rice — (D) *Ephelis oryzae*
(v) Blast disease of rice — (E) *Ustilaginoidea virens*

Ans: (i)—( ), (ii)—( ), (iii)—( ), (iv)—( ), (v)—( )

2. (i) Anthracnose stalk rot of maize — (A) Stalk and rind are dark-brown, water-soaked lesions; soft, slimy tissue; foul odor; twists during lodging.
(ii) Charcoal Rot of maize — (B) Pith and Rind are soft, brown, and water–soaked; twisted stalk during lodging.
(iii) Gibberella Stalk Rot of maize — (C) Disintegrated pith and covered in silver-black fungal structures ("dust"); stalk and roots covered in black fungal structures.
(iv) Bacterial stalk rot of maize — (D) Disintegrated pith with pink-red discolouration; small, black fungal structures (perithecia) on the stalk.
(v) Pythium Stalk Rot of maize — (E) Stalk black, shiny lesions, often blotchy in appearance; disintegrated pith.

Ans: (i)—( ), (ii)—( ), (iii)—( ), (iv)—( ), (v)—( )

3. (i) Sunken dead spots — (A) Chlorosis
(ii) Tissue browning — (B) Blight
(iii) Dusty mass of spores on the host — (C) Necrosis
(iv) Cell death — (D) Anthracnose
(v) Chlorophyll degradation — (E) Smut

Ans: (i)—( ), (ii)—( ), (iii)—( ), (iv)—( ), (v)—( )

4. (i) Bakanae disease of rice — (A) *Sclerophthora rayssiae* var. *zeae*
(ii) Ergot of bajara — (B) *Ephelis oryzae*

(iii) Brown stripe downy mildew of maize	(C) *Cercosporidium personatum*
(iv) Late leaf spot of groundnut	(D) *Claviceps microcephala*
(v) False ergot of rice	(E) *Fusarium fujikuroi*

Ans: (i)—( ), (ii)—( ), (iii)—( ), (iv)—( ), (v)—( )

5. (i) Kalahasti malady of groundnut	(A) *Entyloma oryzae*
(ii) Downy mildew of pearl millet	(B) *Ustilago maydis*
(iii) Brown spot of rice	(C) *Tylenchorhynchus brevelineatus*
(iv) Common smut of maize	(D) *Bipolaris oryzae*
(v) Leaf smut of rice	(E) *Sclerospora graminicola*

Ans: (i)—( ), (ii)—( ), (iii)—( ), (iv)—( ), (v)—( )

6. (i) Tikka disease	(A) Tomato
(ii) Tungro disease	(B) Groundnut
(iii) Black arm	(C) Rice
(iv) Buckeye rot	(D) Tobacco
(v) Wildfire	(E) Cotton

Ans: (i)—( ), (ii)—( ), (iii)—( ), (iv)—( ), (v)—( )

7. (i) White bud of maize	(A) Molybdenum deficiency
(ii) Blossom end rot of tomato	(B) Oxygen deficiency
(iii) Black-heart of potato	(C) Nitrogen deficiency
(iv) Whiptail of cauliflower	(D) Calcium deficiency
(v) Red leaf of cotton	(E) Zinc deficiency

Ans: (i)—( ), (ii)—( ), (iii)—( ), (iv)—( ), (v)—( )

8. (i) Blast of rice	(A) *Aphelenchoides besseyi*
(ii) False smut of rice	(B) *Pyricularia oryzae*
(iii) White-tip of rice	(C) *Fusarium fujikuroi*
(iv) Stem rot of rice	(D) *Ustilaginoidea virens*
(v) Bakanae disease of rice	(E) *Sclerotium oryzae*

Ans: (i)—( ), (ii)—( ), (iii)—( ), (iv)—( ), (v)—( )

9. (i) Covered smut of sorghum	(A) *Gibberella fujikuroi*
(ii) Head Smut of sorghum	(B) *Sporisorium ehrenbergii*
(iii) Long smut of sorghum	(C) *Sporisorium cruentum*
(iv) Loose kernel smut of sorghum	(D) *Sporisorium reilianum*
(v) Pokkah boeng of sorghum	(E) *Sporisorium sorghi*

Ans: (i)—( ), (ii)—( ), (iii)—( ), (iv)—( ), (v)—( )

10. (i) Witchweed	(A) Sugarbeet
(ii) Dodder	(B) Mango
(iii) Broomrape	(C) Sorghum

(iv)	Giant mistletoe	(D)	Cotton
(v)	Red leaf	(E)	Tobacco

Ans: (i)—( ), (ii)—( ), (iii)—( ), (iv)—( ), (v)—( )

11.

(i)	Exanthema in citrus	(A)	Calcium
(ii)	Whiptail of cauliflower	(B)	Zinc
(iii)	Blossom end rot of tomato	(C)	Copper
(iv)	Khaira disease of rice	(D)	Molybdenum
(v)	Pahala disease of sugarcane	(E)	Manganese

Ans: (i)—( ), (ii)—( ), (iii)—( ), (iv)—( ), (v)—( )

12.

(i)	Green ear disease of maize	(A)	*Tolyposporium penicillariae*
(ii)	Twisted top of sorghum	(B)	*Macrophomina phaseolina*
(iii)	Smut of bajara	(C)	*Gibberella fujikuroi*
(iv)	Dry root rot of groundnut	(D)	*Rhizoctonia solani*
(v)	Banded leaf and sheath blight of maize	(E)	*Sclerospora graminicola*

Ans: (i)—( ), (ii)—( ), (iii)—( ), (iv)—( ), (v)—( )

13.

(i)	Testa nematode	(A)	*Phytophthora parasitica*
(ii)	Pigeon pea wilt	(B)	*Aphelenchoides arachidis*
(iii)	Soybean cyst nematode	(C)	*Rhizoctonia solani*
(iv)	Web blight of soybean	(D)	*Heterodera glycines*
(v)	Seedling blight of caster	(E)	*Fusarium udum* f. sp. *cajani*

Ans: (i)—( ), (ii)—( ), (iii)—( ), (iv)—( ), (v)—( )

14.

(i)	Bacterial blight of soybean	(A)	*Pseudomonas syringae* pv. *tabaci*
(ii)	Bacterial pustule of soybean	(B)	*Cercospora sojina*
(iii)	Bacterial tan spot of soybean	(C)	*Pseudomonas savastanoi* pv. *glycinea*
(iv)	Frogeye leaf spot of soybean	(D)	*Xanthomonas axonopodis* pv. *glycines*
(v)	Wildfire of soybean	(E)	*Curtobacterium flaccumfaciens* pv. *flaccumfaciens*

Ans: (i)—( ), (ii)—( ), (iii)—( ), (iv)—( ), (v)—( )

15.

(i)	Seedling blight of caster	(A)	*Rhizoctonia solani*
(ii)	Black shank and leaf blight of tobacco	(B)	*Phytophthora parasitica*
(iii)	Blight of pigeon pea	(C)	*Phytophthora parasitica* var. *nicotianae*
(iv)	Guava wilt	(D)	*Phytopthora drechsleri* f. sp. *cajani*
(v)	Sore shin and damping-off of tobacco	(E)	*Fusarium oxysporum* f. sp. *psidii*

Ans: (i)—( ), (ii)—( ), (iii)—( ), (iv)—( ), (v)—( )

16.

(i)	Smut	(A)	Virus
(ii)	Mosaic	(B)	Bacteria
(iii)	Root knot	(C)	Nematode

(iv) Canker | (D) Fungus
(v) Red rust | (E) Algae

Ans: (i)—( ), (ii)—( ), (iii)—( ), (iv)—( ), (v)—( )

17. (i) Wheat rust | (A) Sclerotia
(ii) Sheath blight of rice | (B) Acervuli
(iii) Bean anthracnose | (C) Ooze
(iv) Fusarium wilt of pigeon pea | (D) Uredospores
(v) Bacterial wilt | (E) Tyloses

Ans: (i)—( ), (ii)—( ), (iii)—( ), (iv)—( ), (v)—( )

18. (i) Black sigatoka of banana | (A) *Colletotrichum musae*
(ii) Cigar-end rot of banana | (B) *Fusarium oxysporum* f. sp. *cubense*
(iii) Panama disease of banana | (C) *Mycosphaerella fijiensis*
(iv) Yellow sigatoka of banana | (D) *Verticillium theobromae*
(v) Anthracnose of banana | (E) *Mycosphaerella musicola*

Ans: (i)—( ), (ii)—( ), (iii)—( ), (iv)—( ), (v)—( )

19. (i) Foot rot of papaya | (A) Virus
(ii) Little leaf of brinjal | (B) *Albugo*
(iii) Ring spot of papaya | (C) *Plasmodiophora*
(iv) Club root of crucifers | (D) *Pythium*
(v) White rust of crucifers | (E) Phytoplasma

Ans: (i)—( ), (ii)—( ), (iii)—( ), (iv)—( ), (v)—( )

20. (i) Host response to infection | (A) Disease
(ii) Pathogen structures | (B) Symptom
(iii) Interaction of host, pathogen, and environment | (C) Necrosis
(iv) Dead tissues | (D) Chlorosis
(v) Degradation of chlorophyll | (E) Sign

Ans: (i)—( ), (ii)—( ), (iii)—( ), (iv)—( ), (v)—( )

21. (i) Red rust of tea | (A) *Exobasidium vexans*
(ii) Blister blight of tea | (B) *Phytophthora palmivora*
(iii) Leaf rust of coffee | (C) *Cephaleuros parasitica*
(iv) Bud rot of coconut | (D) *Phytophthora colocasiae*
(v) Taro blight | (E) *Hemileia vastatrix*

Ans: (i)—( ), (ii)—( ), (iii)—( ), (iv)—( ), (v)—( )

22. (i) Grey leaf spot of coconut | (A) *Ganoderma* spp.
(ii) Root wilt of coconut | (B) *Ceratocystis paradoxa*
(iii) Stem bleeding of coconut | (C) Phytoplasma
(iv) Basal stem rot (BSR) of coconut | (D) *Pestalotiopsis palmarum*
(v) Bud rot of coconut | (E) *Phytophthora palmivora*

Ans: (i)—( ), (ii)—( ), (iii)—( ), (iv)—( ), (v)—( )

23. (i) Rice hoja blanca (A) Nematode
(ii) Rice white-tip (B) Fungus
(iii) Khaira disease of rice (C) Bacteria
(iv) Panicle blight of rice (D) Virus
(v) Sheath blight of rice (E) Zinc deficiency

Ans: (i)—( ), (ii)—( ), (iii)—( ), (iv)—( ), (v)—( )

24. (i) Virus (A) Wilt
(ii) Phytoplasma (B) Damping off
(iii) *Fusarium* (C) Canker
(iv) *Pythium* (D) Little leaf
(v) *Xanthomonas* (E) Mosaic

Ans: (i)—( ), (ii)—( ), (iii)—( ), (iv)—( ), (v)—( )

25. (i) Powdery mildew (A) *Puccinia*
(ii) Downy mildew (B) *Fusarium*
(iii) Rust (C) *Sclerospora*
(iv) Wilt (D) *Phytophthora*
(v) Blight (E) *Erysiphe*

Ans: (i)—( ), (ii)—( ), (iii)—( ), (iv)—( ), (v)—( )

26. (i) Sesamum phyllody (A) Aphid
(ii) Pigeon pea sterility mosaic (B) Leaf hopper
(iii) Wheat streak mosaic (C) Eriophyid mite
(iv) Bhindi yellow vein mosaic (D) Curl mite
(v) Banana bunchy top (E) White fly

Ans: (i)—( ), (ii)—( ), (iii)—( ), (iv)—( ), (v)—( )

Answers: Multiples Choice Questions (MCQs)

1. a **2.** d **3.** d **4.** a **5.** a **6.** d **7.** a **8.** b **9.** b **10.** c
11. a **12.** b **13.** a **14.** d **15.** a **16.** b **17.** b **18.** d **19.** d **20.** d
21. d **22.** b **23.** c **24.** d **25.** b **26.** d **27.** d **28.** a **29.** d **30.** c
31. c **32.** a **33.** b **34.** d **35.** d **36.** a **37.** d **38.** d **39** d **40.** b
41. d **42.** d **43.** d **44.** b **45.** b **46.** b **47.** d **48.** b **49.** d **50.** d
51. d **52.** c **53.** d **54.** b **55.** d **56.** d **57.** c **58.** d **59.** a **60.** d
61. a **62.** d **63.** d **64.** b **65.** d **66.** a **67.** b **68.** a **69.** d **70.** c
71. d **72.** a **73.** b **74.** d **75.** a **76.** a **77.** d **78.** a **79.** c **80.** c
81. a **82.** d **83.** a **84.** d **85.** b **86.** b **87.** a **88.** a **89.** a **90.** d
91. d **92.** c **93.** d **94.** d **95.** b **96.** c **97.** c **98.** c **99.** d **100.** d
101. d **102.** b **103.** b **104.** c **105.** a **106.** b **107.** c **108.** b **109.** d **110.** b
111. d **112.** b **113.** d **114.** d **115.** d **116.** d **117.** d **118.** d **119.** d **120.** a
121. b **122.** b **123.** d **124.** d **125.** c **126.** b **127.** d **128.** a **129.** d **130.** d

131. d 132. d 133. d 134. a 135. c 136. d 137. d 138. b 139. d 140. d
141. a 142. c 143. a 144. d 145. d 146. a 147. d 148. d 149. d 150. d
151. d 152. b 153. d 154. c 155. b 156. d 157. d 158. d 159. d 160. d
161. d 162. a 163. d 164. d 165. a 166. a 167. d 168. b 169. d 170. d
171. a 172. b 173. c 174. b 175. a 176. d 177. d 178. b 179. d 180. d
181. b 182. b 183. d 184. b 185. d 186. d 187. a 188. b 189. a 190. d
191. d 192. c 193. a 194. d 195. a 196. b 197. d 198. d 199. a 200. d
201. c 202. b 203. d 204. d 205. a 206. b 207. c 208. d 209. a 210. b
211. b 212. b 213. b 214. b 215. d 216. a 217. a 218. d 219. c 220. a
221. b 222. b 223. c 224. d 225. a 226. d 227. d 228. d 229. d 230. a
231. d 232. d 233. a 234. b 235. b 236. d 237. d 238. b 239. a 240. d
241. d 242. a 243. d 244. d 245. a 246. a 247. b 248. d 249. d 250. c
251. d 252. a 253. b 254. d 255. b 256. c 257. d 258. d 259. d 260. d
261. a 262. a 263. b 264. c 265. d 266. d 267. a 268. d 269. d 270. b
271. c 272. a 273. a 274. a 275. a 276. a 277. b 278. b 279. a 280. d
281. d 282. a 283. d 284. d 285. d 286. d 287. d 288. d 289. d 290. b
291. d 292. d 293. d 294. d 295. b 296. b 297. b 298. d 299. d 300. b
301. d 302. a 303. d 304. c 305. a 306. d 307. a 308. d 309. d 310. b
311. a 312. d 313. a 314. d 315. d 316. c 317. b 318. a 319. a 320. d
321. d 322. b 323. d 324. d 325. b 326. b 327. d 328. a 329. d 330. d
331. d 332. d 333. b 334. d 335. b 336. b 337. b 338. d 339. c 340. b
341. b 342. d 343. b 344. d 345. a 346. d 347. a 348. c 349. d 350. a
351. d 352. d 353. b 354. b 355. b 356. b 357. b 358. c 359. c 360. b
361. b 362. b 363. a 364. a 365. b 366. a 367. d 368. c 369. a 370. a
371. d 372. d 373. b 374. a 375. b 376. d 377. d 378. b 379. b 380. a
381. c 382. b 383. b 384. c 385. b 386. b 387. c 388. d 389. d 390. a
391. d 392. b 393. b 394. a 395. a 396. a 397. a 398. b 399. b 400. b
401. b 402. b 403. d 404. c 405. d 406. a 407. c 408. a 409. c 410. d
411. d 412. d 413. a 414. d 415. a 416. a 417. b 418. c 419. b 420. a
421. c 422. a 423. a 424. a 425. a 426. d 427. a 428. a 429. b 430. a
431. b 432. b 433. b 434. c 435. b 436. b 437. a 438. a 439. a 440. a
441. b 442. b 443. d 444. d 445. d 446. c 447. d 448. d 449. d 450. b
451. b 452. a 453. d 454. b 455. c 456. c 457. a 458. d 459. a 460. a
461. c 462. b 463. b 464. c 465. b 466. a 467. c 468. c 469. d 470. b
471. b 472. d 473. d 474. c 475. a 476. d 477. d 478. d 479. b 480. d
481. a 482. c 483. d 484. d 485. d 486. b 487. d 488. a 489. a 490. a
491. a 492. d 493. d 494. c 495. d 496. d 497. b 498. c 499. b 500. d
501. d 502. c 503. c 504. c 505. b 506. d

Answers: Match the Following

1. (i)—(E), (ii)—(C), (iii)—(D), (iv)—(B), (v)—(A)
2. (i)—(E), (ii)—(C), (iii)—(D), (iv)—(A), (v)—(B)
3. (i)—(D), (ii)—(B), (iii)—(E), (iv)—(C), (v)—(A)
4. (i)—(E), (ii)—(D), (iii)—(A), (iv)—(C), (v)—(B)
5. (i)—(C), (ii)—(E), (iii)—(D), (iv)—(B), (v)—(A)
6. (i)—(B), (ii)—(C), (iii)—(E), (iv)—(A), (v)—(D)
7. (i)—(E), (ii)—(D), (iii)—(B), (iv)—(A), (v)—(C)
8. (i)—(B), (ii)—(D), (iii)—(A), (iv)—(E), (v)—(C)
9. (i)—(E), (ii)—(D), (iii)—(B), (iv)—(C), (v)—(A)
10. (i)—(C), (ii)—(A), (iii)—(E), (iv)—(B), (v)—(D)
11. (i)—(C), (ii)—(D), (iii)—(A), (iv)—(B), (v)—(E)
12. (i)—(E), (ii)—(C), (iii)—(A), (iv)—(B), (v)—(D)
13. (i)—(B), (ii)—(E), (iii)—(D), (iv)—(C), (v)—(A)
14. (i)—(C), (ii)—(D), (iii)—(E), (iv)—(B), (v)—(A)
15. (i)—(B), (ii)—(C), (iii)—(D), (iv)—(E), (v)—(A)
16. (i)—(D), (ii)—(A), (iii)—(C), (iv)—(B), (v)—(E)
17. (i)—(D), (ii)—(A), (iii)—(B), (iv)—(E), (v)—(C)
18. (i)—(C), (ii)—(D), (iii)—(B), (iv)—(E), (v)—(A)
19. (i)—(D), (ii)—(E), (iii)—(A), (iv)—(C), (v)—(B)
20. (i)—(B), (ii)—(E), (iii)—(A), (iv)—(C), (v)—(D)
21. (i)—(C), (ii)—(A), (iii)—(E), (iv)—(B), (v)—(D)
22. (i)—(D), (ii)—(C), (iii)—(B), (iv)—(A), (v)—(E)
23. (i)—(D), (ii)—(A), (iii)—(E), (iv)—(C), (v)—(B)
24. (i)—(E), (ii)—(D), (iii)—(A), (iv)—(B), (v)—(C)
25. (i)—(E), (ii)—(C), (iii)—(A), (iv)—(B), (v)—(D)
26. (i)—(B), (ii)—(C), (iii)—(D), (iv)—(E), (v)—(A)

References

Books

Aglave, B. 2019. *Handbook of Plant Disease Identification and Management*. CRC Press and Disease Management. Taylor & Francis Group, pp. 751.

Awasthi, L.P. 2015. *Recent Advances in the Diagnosis and Management of Plant Diseases*. Springer India, pp. 285.

Chakrabarti, D.K. 2011. *Mango Malformation*. Springer-Science + Business Media B.V., pp. 147.

Chand, G. and Kumar, S. 2016. *Crop Diseases and Their Management: Integrated Approaches*. Apple Academic Press Inc., pp. 385.

Chaube, H.S. and Pundhir, V.S. 2012. *Crop Diseases and Their Management*. PHI Learning, pp. 703.

Cooke, T., Persley, D. and House, S. 2009. *Diseases of Fruit Crops in Australia*. CSIRO Publishing, pp. 269.

Gatak, A. and Ansar, M. 2020. *The Vegetable Pathosystem: Ecology, Disease Mechanism, and Management*. Apple Academic Press Inc., pp. 527.

Horst, R.K. 2001. *Westcott's Plant Disease Handbook*. 6th ed., Kluwer Academic Publishers, pp. 733.

Husaini, A.M. and Ner, D. 2016. *Strawberry: Growth, Development and Diseases*. CABI. pp. 313.

Loebenstein, G., Cemer, V. and Thottappill, G. 2003. *Virus and Virus-like Diseases of Major Crops in Developing Countries*. Springer-Science + Business Media, B.V., pp. 791.

Mukerji, K.G. 2004. *Disease Management of Fruits and Vegetables,* Vol. 1, Kluwer Academic Publishers, pp. 553.

Nair, K.P.P. 2013. *The Agronomy and Economy of Turmeric and Ginger: The Invaluable Medicinal Spice Crops*. Elsevier Inc., pp. 513.

Nyvall, R.F. 1999. *Field Crop Diseases.* 3rd ed., Iowa State University Press. pp. 969.

Ploetz, R.C. 2003. *Diseases of Tropical Fruit Crops*. CABI Publishing, pp. 499.

Saharan, G.S. and Mehta, N. 2008. *Sclerotinia Diseases of Crop Plants: Biology, Ecology*. Springer-Science + Business Media, B.V., pp. 481.

Saharan, G.S. Mehta, N. and Meena, P.D. 2016. *Alternaria Diseases of Crucifers: Biology, Ecology and Disease Management*. Springer-Science + Business Media, Singapore, pp 293.

Singh, R.S. 2018. *Plant Diseases.* 10th ed., Medtech Science Press, pp. 821.

Srivastava, J.N. and Singh, A.K. 2021. *Diseases of Field Crops: Diagnosis and Management: Cereals, Small Millets, and Fiber Crops*. Vol. 1, Apple Academic Press Inc. pp. 387.

Srivastava, J.N. and Singh, A.K. 2021. *Diseases of Field Crops: Diagnosis and Management: Pulses, Oil Seeds, Narcotics, and Sugar Crops*. Vol. 1, Apple Academic Press Inc., pp. 365.

Tennant, P. and Fermin, G. 2015. *Virus Diseases of Tropical and Subtropical Crops*. CABI Publishing, pp. 241.

Thind, B.S. 2020. *Phytopathogenic Bacteria and Plant Diseases*. CRC Press Taylor & Francis Group, pp. 367.

Walia, R.K. and Bajaj, H.K. 2014. *Textbook of Introductory Plant Nematology*. Directorate of Information and Publication of Agriculture, ICAR, New Delhi, pp. 228.

Waller, J.M., Bigger M. and Hillocks, R.J. 2007. *Coffee Pests, Diseases and Their Management*. CABI Publishing, pp. 423.

Research/Review Articles/Technical Bulletins

Ahuja, D.B. and Chattopadhyay, C. 2015. Pests of fruit trees (citrus, banana, mango, pomegranate and sapota): E-pest surveillance and pest management advisory. ICAR-National Research Centre for Integrated Pest Management, New Delhi and State Department of Horticulture, Commissionerate of Agriculture, Pune (Maharashtra), pp. 124.

Ajayi-Oyetundea, O.O. and Bradley, C.A. 2018. *Rhizoctonia solani*: taxonomy, population biology, and management of Rhizoctonia seedling disease of soybean. *Plant Pathology*, 67, 3–17.

Arnold, D.L., Helen, C.L., Jackson, R.W. and Mansfield, J.W. 2011. *Pseudomonas syringae* pv. *phaseolicola*: from 'has bean' to supermodel. *Molecular Plant Pathology*, 12(7): 617–627.

Bashyal, B.M. 2018. Etiology of an emerging disease: bakanae of rice. *Indian Phytopathology*, 71: 485–494.

Blomme, G., Dita, M., Jacobsen, K.S., Pérez, et al. 2017. Bacterial diseases of bananas and enset: Current state of knowledge and integrated approaches toward sustainable management. *Front. Plant Sci.* 8: 1290.

Chitra, N., Suhas, Y., Malathi, S., et al. 2011. Manual for pigeonpea pest surveillance: National Initiative on Climate Resilient Agriculture. ICAR-NCIPM, New Dehli, pp. 29.

CIAT and FAO. 2015. Current status of moko disease and black sigatoka in Latin America and the Caribbean, and options for managing them. pp. 32.

Dubrow, Z.E. and Bogdanove, A.J. 2021. Genomic insights advance the fight against black rot of crucifers. *Journal of General Plant Pathology*, 87: 127–136.

FAO. 2007. Regional vegetable IPM programme green bean ecological guide. pp. 78.

Forghani, F. and Hajihassani, A. 2020. Recent advances in the development of environmentally benign treatments to control root-knot nematodes. *Front. Plant Sci.* 11: 1125. doi: 10.3389/fpls.2020.01125

Gonza'lez, A.J., Landeras, E. and Mendoza, M.C. 2000. Pathovars of *Pseudomonas syringae* causing bacterial brown spot and halo blight in *Phaseolus vulgaris* L. are distinguishable by ribotyping. *Applied and Environmental Microbiology*, 66(2): 850–854.

Gupta, A.K., Solanki, I.S., Bashyal, B.M., et al. 2015. Bakanae of rice: an emerging disease in Asia. *The Journal of Animal & Plant Sciences*, 25(6): 1499–1514.

Hajimorad, M.R., Domier, L.L., Tolin, S.A., et al. 2018. *Soybean mosaic virus*: a successful potyvirus with a wide distribution but restricted natural host range. *Molecular Plant Pathology*, 19(7): 1563–1579.

Jagdale, S., Rao, U. and Giri, A.P. 2021. Effectors of root-knot nematodes: An arsenal for successful parasitism. *Front. Plant Sci.* 12: 800030. doi: 10.3389/fpls.2021.800030

Jing, L.F. and Suga, H. 2021. Various methods for controlling the bakanae disease in rice. *Reviews in Agricultural Science*, 9: 195–205.

Kharayat, B.S. and Singh, Y. 2020. Ralstonia-Tomato Pathosystem: Pathogen biology, host-pathogen interaction, and management. *In*: Vegetable pathosystem. (Editor: Abhijeet Gatak and Mohammad Ansar). Apple Academic Press Inc, pp. 85–124.

Kim, M., Shim, C., Lee, J. and Wangchuk, C. 2022. Hot water treatment as seed disinfection techniques for organic and eco-friendly environmental agricultural crop cultivation. *Agriculture*, 12, 1081.

Kumar, P.L. 2009. Virus detection in banana a laboratory manual. IITA, pp. 50.

Le, D.P., Smith, M., Hudler, G.W. and Aitken, E. 2014. Pythium soft rot of ginger: Detection and identification of the causal pathogens, and their control. *Crop Protection*, 65: 153 167.

Lizardo, R.C.M., Pinili, M.S., Diaz, M.G.Q. and Cumagun, C.J.R. 2022. Screening for Resistance in selected tomato varieties against the root-knot nematode *Meloidogyne incognita* in the Philippines using a molecular marker and biochemical analysis. *Plants*, 11, 1354.

Naimuddin, Akram, Md. and Singh, N.P. 2016. Yellow mosaic of mungbean and urdbean: current status and future strategies. *Journal of Food Legumes*, 29(2): 77–93.

NICRA team of Tomato Pest Surveillance. 2012. *Manual for Tomato Pest Surveillance*. Jointly published by NCIPM, New Delhi, and CRIDA, Hyderabad, IIHR, Bengaluru and IIVR,Varanasi, pp. 39.

Noble, T.J., Young, A.J., Douglas, C.A., et al. 2019. Diagnosis and management of halo blight in Australian mungbeans: A review. *Crop & Pasture Science*, CSIRO Publishing, 70, 195–203.

Pande, S., Sharma, M. and Guvvala, G. 2013. An updated review of biology, pathogenicity, epidemiology and management of wilt disease of pigeonpea (*Cajanus cajan* (L.) Millsp.). *Journal of Food Legumes*, 26(1 and 2): 1–14.

Pande, S., Sharma, M., Naga Mangla, U., et al. 2011. Phytophthora blight of pigeonpea [*Cajanus cajan* (L.) Millsp.]: An updating review of biology, pathogenicity, and disease management. *Crop Protection*, 30, 951–957.

Pandey, A.K., Burlakoti, R.R., Kenyon, L., et al. 2018. Perspectives and challenges for sustainable management of fungal diseases of mungbean [*Vigna radiata* (L.) R. Wilczek var. *radiata*]: A review. *Front. Environ. Sci.* 6: 53.

Plant Health Australia. 2006. *Bacterial wilt of Banana Diagnostics Manual*, pp. 58.

Prakesh, A., Bentur, J.S., Srinivas, P.M., et al. 2014. Integrated pest management for rice, pp. 43.

Rawat, K., Tripathi, S.B., Kaushik, N. and Bashyal, B.M. 2022. Management of bakanae disease of rice using biocontrol agents and insights into their biocontrol mechanisms. *Arch. Microbiol.*, 204, 401. https://doi.org/10.1007/s00203-022-02999-3

Rehman, F.Ur, Kalsoom, M., Adnan, Naz, M., et al. 2021. Soybean mosaic disease (SMD): A review. *Egyptian Journal of Basic and Applied Sciences*, 8: 1, 12–16.

Seid, A., Fininsa, C., Mekete, T., Decraemer, W., et al. 2015. Tomato (*Solanum lycopersicum*) and root-knot nematodes (*Meloidogyne* spp.): a century-old battle. *Nematology*, 17, 995–1009.

Sen, S., Rai, M., Das, D., Chandra, S., et al. 2020. Blister blight a threatened problem in tea industry: A review. *Journal of King Saud University–Science*, 32, 3265–3272.

Sharma, O.P., Bambawale, O.M., Gopali, J.B., et al. 2011. *Field Guide Mungbean and Urdbean*. ICAR-NCIPM, pp. 40.

Sharma, O.P., Gopali, J.B., Yelshetty, S., et al. 2010. *Pests of Pigeon Pea and Their Management*. ICAR-NCIPM, LBS Building, IARI Campus, New Delhi-110012, India, pp. 92.

Shetty, H.S., Raj Niranjan, S., Kini, et al. 2016. Downy mildew of pearl millet and its management. All India Coordinated Research Project on Pearl Millet (Indian Council of Agricultural Research), Mandor, Jodhpur–342304, pp. 53.

Singh, D., Jackson, G., Hunter, D., et al. 2012. Taro leaf blight-a threat to food security. *Agriculture*, 2: 182–203.

Singh, Y., Sharma, D. and Kharayat, B.S. 2021. Major Diseases of Sorghum and Their Management. *In*: *Diseases of Field Crops: Diagnosis and Management*, Volume 1: Cereals, Small Millets, and Fiber Crops, Srivastava, J.N. and A.K. Singh (Eds.), Apple Academic Press. Inc., USA, p. 153–182.

Snehalatharani, A., Maheswarappa, H.P., Devappa, V., et al. 2016. Status of coconut basal stem rot disease in India–A review. *Indian Journal of Agricultural Sciences*, 86 (12): 1519–1529.

Srinivasulu, B., Sujatha, A., Kalpana, M., et al. 2008. Biocontrol of Ganoderma wilt (basal stem rot) disease of coconut. AICRP on palms (ICAR) Andhra Pradesh Horticultural University Horticultural Research Station, Ambajipeta-533214, E.G. Dist., A.P., pp. 30.

Sun, S., Zhi, Y., Zhu, Z., et al. 2017. An emerging disease caused by *Pseudomonas syringae* pv. *phaseolicola* threatens mungbean production in China. *Plant Disease*, 101: 95–102.

Talhinhas, P., Batista, D., Diniz, I., et al. 2017. The coffee leaf rust pathogen *Hemileia vastatrix*: one and a half centuries around the tropics. *Molecular Plant Pathology*, 18(8): 1039–1051.

Thakur, R.P., Sharma, R. and Rao, V.P. 2011. Screening techniques for pearl millet diseases. *Information Bulletin No. 89.* Patancheru-502324, Andhra Pradesh, India, International Crops Research Institute for the Semi-Arid Tropics, pp. 56.

Torres, G.A., Sarria, G.A., Martinez, G., et al. 2016. Bud rot caused by *Phytophthora palmivora*: A destructive emerging disease of oil palm. *Phytopathology*, 106: 320–329.

Tripathi, L., Mwangi, M., Abele, S., et al. 2009. Bacterial wilt: A threat to banana production in east and central Africa. *Plant Disease*, 93(5): 440–451.

Uwamahoro, F., Berlin, A., Bylund, H., et al. 2019. Management strategies for banana Xanthomonas wilt in Rwanda include mixing indigenous and improved cultivars. *Agronomy for Sustainable Development*, 39: 22.

Vicente, J.G. and Holub, E.B. 2013. *Xanthomonas campestris* pv. *campestris* (cause of black rot of crucifers) in the genomic era is still a worldwide threat to brassica crops. *Molecular Plant Pathology*, 14(1): 2–18.

Viljoen, A., Mahuku, G., Massawe, C., et al. 2016. Banana pests and diseases: Field guide for disease diagnostics and data collection. IITA, Ibadan, Nigeria, pp. 96.

Virginio Filho, E.de.M. and Domian, C.A. 2019. Prevention and control of coffee leaf rust handbook of best practices for extension agents and facilitators. *Technical Manual,* No. 131. Tropical Agricultural Research and Higher Education Center (CATIE), pp. 95.

Widyasari, K., Alazem, M. and Kim, K.H. 2020. Soybean resistance to *Soybean mosaic virus.* Plants, 9, 219.

Wilson, J.P. 1999. Pearl Millet Diseases: A compilation of information on the known pathogens of pearl millet, *Pennisetum glaucum* (L.) R. Br. USDA, ARS, *Agriculture Handbook,* No. 716.

Index